Early Village Farming Culture

with special reference to

Eastern and North Eastern India

Early Village Farming Culture
with special reference to
Eastern and North Eastern India

Mukta Raut Dey

2011
Kaveri Books
New Delhi - 110 002

First Published in 2011
ISBN-13 : 978-81-7479-113-9

Kaveri Books
4832/24, Ansari Road
New Delhi - 110 002 (India)
Tel.: 011-2328 8140, 2324 5799
E-mail: kaveribooks@vsnl.com
website : www.kaveribooks.com

Laser Type Setting by
Aarati Computers, Delhi - 110 009.

Printed at
G.S. Offset Printers, Delhi-110 032.

PRINTED IN INDIA

Foreword

Neolithic stage of India *vis-a-vis* of East and North East India has not been so far given a complete documentation. In this respect Dr. Mukta Raut Dey has made an attempt at reconstructing the early village farming cultures of East and North East India in a well researched manner. In fact this is an outcome of her Ph. D. work. Neolithic in Indian context is mainly substantiated by secondary data, such as, pottery, tool types and their production technology in most of the regions. Excepting Ganga valley, hardly anywhere the Neolithic sites have yielded remains of cultivated cereals. In Eastern India the sites, which yielded remains of rice are quite late in date. Several attempts had been made earlier by the Botanists as well as Anthropologists to point out the place of origin of rice in eastern India in the Jeypur tracts of the state of Orissa. This was done mainly on the basis of preponderance of wild varieties of rice present in this area and the presence of species of *Oryza,* which could be ancestral to the cultivated species *Oryza sativa*.

In both the states of East and North East India rice is the staple crop. This area is witness to diverse system of cropping. It ranges from shifting, terrace and to flat land cultivation. Each system has got its own mechanism and technique. Though shifting cultivation is considered as a primitive one, it involves elaborate understanding of the environmental potentiality. Terrace cultivation in the North east is a phenomenon in itself. The example of the Apatanis may be given. This tribe of Arunachal Pradesh has excelled in terrace cultivation.

The words tribe, indigenous and primitive people are politically loaded and invites controversy. However, due to isolation and long time association with the local environment the so called tribal people have retained a store house of information about the plants and animals they encounter in their daily life. This often goes long back in time. Their socio-cultural pattern also conforms to this. It has been found that these people even follow some customary law to preserve their environment, cultigens and traditional mode of cultivation.

Ethno archaeology as a tool for reconstruction of prehistoric ways of life attained its importance in the last century, especially with the advent of concepts associated with New World Archaeology. The method is not at all free from errors. Space and time are major constraints involving the methodology.

Dr. Mukta Raut Dey has applied the ethno-archaeological approach for the reconstruction of the ways of life of the early village farming communities of the East and North East India. She had utilized the information available in the ethnographic accounts of the tribes residing in the area. To this she has added archeological evidences found from the Neolithic village sites. A comparison is made of the artifacts found in Neolithic sites with the analogous types used by the tribes in the areas. Not only has she tried to reconstruct the functions of the implements but she has focused on the probable method of cultivation of the early farmers of the area. She has also tried to reconstruct the house types and other relevant features of early village farming culture on the basis of data available from the present day tribal residents of the region. All along she is very cautious about the fallacy that may come up with the application of the methodology. The book will provide a wealth of information on the topic both to the lay readers and to the schlars involved in similar studies.

14th April 2010

Prof. Ranjana Roy
Former Professor, Emeritus (UGC),
Anthropology, Calcutta University

Acknowledgement

In the beginning, I would like to express my sincere gratitude to late Prof. A. C. Pal, Prof D.K Bhattacharya and Dr. R. K. Chattopadhyay and Prof. Chitrarekha Gupta for their avid guidance in completing this work.

My sincere thanks are also due to Prof. Ranjana Ray, Calcutta University, Dr. V. R. Rao, Director, Anthropological Survey of India, Shri S. B. Ota, Archaeological Survey of India for all possible help and co-operation.

I bestow my gratitude to Dr. S. S. Biswas, former DG National Museum, Mr. K. N. Dikshit, former Jt. DG Archaeological Survey of India for encouraging me to prepare the manuscript for publication. I want to thank officials of Anthropological Survey of India for providing me relevant photographs, Mr. Bharat Singh, Indian Archaeological Society, Mr. Umesh Mande, Archaeological Survey of India for helping in the preparation of plates. Maps are produced by Shri N. C. Das, Calcutta University and Dr. P. K. Guha, Anthropological Survey of India. I am indebted to them. Last but not the least all my family members specially my father and husband deserve my heartfelt thanks for every possible support, during these years.

— **Mukta Raut Dey**

Preface

The term Early Village Farming encompasses the beginning of farming to the emergence of metal. Here, in this dissertation, the earliest phase of this Early Village Farming Culture which is popularly known as 'Neolithic' culture would be dealt with and I would like to adhere to this most accepted term 'Neolithic' hereafter, in this book. The term 'Neolithic' could be interpreted in two ways – traditionally it is symbolized by the occurrence of ground and polished stone tools, celt, ring stone, bolas, hand made pottery as well as mud houses with domesticated economy. The other facet of the said culture refers to economy as well as issues of social formation which is not often represented in archaeologically retrieved data and one has to look at possibilities of work logistics and the like that is a prerequisite of a labour intensive economy.

An effective village-farming way of life had been achieved in certain portions of South-Western Asia by at least 7000 B.C.E. The level of nascent food producing is best reflected in the archaeological materials of the Natufian group in Palestine-Syro-Lebanese littoral and parts of its hinterland, and by the Karim Shahir group in Iraqi and Iranian Kurdistan. In India, knowledge of the development stages that lead to an early village farming settlement is still not complete. From the Ganga and Belan valley excellent evidences are coming out, which pertain to a well-advanced phase of the effective village-farming-community level, but unfortunately the region under consideration here *ie* eastern and north eastern India, the data for such reconstruction is not comprehensive. With meager archaeological evidence in hand the present venture is an honest attempt to give a better shape to the Neolithic culture of eastern and north eastern India, comprising the states of West Bengal, Bihar, Jharkhand, Orissa, Assam, Meghalaya, Arunachal Pradesh, Manipur, Mizoram, Tripura and Nagaland with ethnographic aid.

Contents

List of Plates

List of Maps & Figures

LIST OF MAPS

LIST OF FIGURES

The illustrations reproduced are from the followings published materials:

Indian Archaeology- A Review

Allchin, B. (1995) *Living Traditions: Studies in Ethnoarchaeology of South Asia,* Oxford and IBH publishing Co.

Ashraf, A. A. (1990) *Prehistoric Arunachal: A Study on Prehistory and Ethnoarchaeology of Kamla Valley*, Directorate of Research, Itanagar.

-- (1994) Anthropo-Archaeological Back Ground of Nishi Material Culture, *BDA*, VIII.

Basa K. K. and P. K. Mohanty (2000) *Archaeology of Orissa,* Vol. I, II, Delhi.

Chakrabarti, Dilip K. (1988) *Theoretical Issues of Indian Archaeology*, Delhi.

-- (1993) *Archaeology of Eastern India: Chhotanagpur Plateau and West Bengal*, Delhi.

Datta, Ashok (1992) *Neolithic Culture in West Bengal,* Delhi.

Goswami, M.C. (1972) 'A Typological Study of Some Prehistoric Tools from Kameng District NEFA', *JASS*, XV (I).

Goswami, M.C. and S.K. Roy (1975) 'A Note on Tangsa Hoe, Arunachal Pradesh', *BDA*, IV.

Haimendorf, C.V.F. (1945) 'The Problem of Megalithic Culture in the Middle India', *MI*, XXXVI (II).

-- (1946) *The Naked Nagas*, Calcutta.

Narayan, B. (1996) *Prehistoric Archaeology of Bihar,* Patna.

Ramesh, N.R. (1986) 'Discovery of Stone Age Tools From Tripura and its Relevance to the Prehistory of South-East Asia', *G.S.M.*, II.

Sankalia, H.D. (1964) *Stone Age Tools*, Poona.

Sen, D (1950) 'A Celt Site in Singhbhum', *MI,* XXX(I).

Sharma, T.C. (1981) 'The Neolithic Pattern of North Eastern India' in M.S. Nagarja Rao (ed) *Madhu: Recent Researches In Indian Archaeology and Art History*, Delhi, p.42.

Singh, B. P. (2004) *Early Farming Communities of the Kaimur,* Vol. I, Jaipur.

Anthropological Survey of India has also provided a number of photographs.

Abbreviations

AA	American Anthropologist.
AGS	American Geographical Society.
ASI	Archaeological Survey of India.
Al	Ancient India.
An Rep. ASI	Annual Report of Anthropological Survery of India.
Bull.	Bulletin.
BCRI	Bull. of Cultural Research Institute.
BDA	Bull. of Department of Anthropology.
BDCRI	Bull. of Deccan College Research Institute.
BSOAS	Bull. of School of Oriental and African Studies.
CA	Current Anthropology.
CB	Chronica Botanica.
COI	Census of India.
DAWB	Directorate of Archaeology, West Bengal
EA	Easteren Anthropologist.
GRI	Geographical Review of India.
GSI	Geological Survey of India.
GSM	Geological Society of Malaysia.
IA	Indian Antiquary.
IAR	Indian Archaeology - A Review.
IF	Indian Folklore.

IMB	Indian Museum Bulletin.
J1	Journal.
JARS	Journal of Assam Research Society.
JASB	Journal of Asiatic Society of Bengal.
JASS	Journal of Assam Science Soceity.
JBORS	Journal of Bihar Orissa Research Society.
JBPP	Journal of Bihar Puravid Parishad.
JIAS	Journal of Indian Anthropological Society.
JOI	Journal of Oriental Institute.
JP	Journal of Palynology.
JPASB	Journal and Proceedings of Asiatic Society of Bengal.
JRAI	Journal of Royal Anthropological Institute.
JRASB	Journal of Royal Asiatic Soceity of Bengal.
JUG	Journal of University of Gauhati.
JWAS	Journal of Washington Academy of Science.
ME	Man and Environment.
MGSI	Memoir of Geological Survery of India.
MI	Man in India.
MUSM.	Museum.
PASB	Proceedings of Asiatic Society of Bengal.
PB	Palaeobotanist.
PIGC	Proceedings of Indian Geological Congress.
PISC	Proceedings of Indian Science Congress.
RASB	Royal Asiatic Society of Bengal.
SC	Science and Culture.
WA	World Archaeology.

Chapter-1

Introduction

The term 'Neolithic' has also its meaning in a general sense have always been prescribed as a phase of significant technological change. In the recent years authors have though emphasized the greater importance of social structure and ideology in determining and perpetuating the basic technological change, yet consideration of Neolithic in India has seldom included discussion on these important issues. It is true that archaeological methods do not provide adequate insight into those activities and thought processes which lead to a specific behaviour yet then these considerations are of great importance and can not be denied. The present study attempts to elucidate the process of development of this momentous change in human cultural history by way of including an anthropological insight along with ecospecific impreratives to understand the various shades of development of Neolithic in India, for instance the preference of *durum* wheat in the west or millet in the south or rice in the east or even adopting slash-and-burn method of cultivation in further east. The variation of these choices can lead to change in work logistics, settlement pattern and consequently also in the ideology. It is needless to emphasize that understanding a structure alone can help us to understand the choice or choices a community makes. The shades in which a community develops its choices for productive economy has largely remained as an epiphenomenon in Indian archaeological discussion. An aim to understand these, hitherto neglected areas of Neolithic culture specifically in eastern and north-eastern regions of India could be undertaken. The entire sub-continent of India has although yielded material remains to understand the rise and growth of food producing economy but there are anomaly of data and interpretation, to some extent.

Under this condition, the basic outline of neolithism and its relationship with human development in India in general and eastern and north-eastern India in particular, is the major concern of present study. In the historiography, we have seen a wide range of discussion about the concept of Neolithic, which may be quoted at the beginning.

Lubbock interpreted Neoliths in terms of its i) association with recent fauna as against extinct fauna of Paleolithic times ii) shaping of the tool by grinding in addition to chipping iii) practice of agriculture and domestication of animal[1].

Childe added few other criteria for this cultural phase *viz*. i) a food producing economy dependent on the domestication of animal and cultivation of plant; hunting and food gathering, however, continues though in a subsidiary role ii) sedentary nature of habitation best exemplified in tells or mounds – sometimes rising to a height of 30 meters or more covering at times several acres iii) absence of dependence on trade for essential commodities though sporadic trade in luxury goods might well be present, as was the case even with some upper Paleolithic and Mesolithic communities, iv) absence of full-time specialist in the society[2].

According to Burkitt, Neolithic must possess features like i) practice of agriculture ii) domestication of animal iii) manufacture of pottery iv) grinding and polishing technique[3].

One of the first serious attempts to study the origin of agriculture by utilizing data from a variety of scientific fields was undertaken by Condolle. He enumerated a number of conditions under which plant cultivation might have been initiated. i) the plants must be productive and easy to rear ii) the climate must not be too rigorous iii) there must be some duration of draught in hot countries iv) there must be some degree of security and settlement v) there must be a pressing necessity arising from insufficient resources from hunting, fishing or gathering of wild plants[4].

C.O. Sauer developed six basic assumptions about the origin of domestication. i) Agriculture does not arise due to food shortage. People on the verge of starvation did not have the leisure time to experiment with improving food plants. ii) the hearth of domestication is found in areas of marked diversification of plants and animals. Regions characterized by both diversified terrain and variable climate, iii) domestication did not begin in the river valleys which are subjected to floods and aridity and require irrigation, iv) the earliest agriculture was practised in "wooded lands", v) the earliest agriculturist had already achieved special skill that predisposed them to domestication, vi) the inventors of agriculture already sedentary, since growing of crops required constant attention. Planting a field then leaving it until harvest time, would have invited predators and lead to loss of harvest[5].

Gordon Childe tried to explain this phenomenon through his '*Oasis propinquity*' theory. He suggested that 'Food production the deliberate cultivation of food plants, especially cereals...was an economic revolution' the condition of incipient desiccation would provide stimulus towards adaptation of food producing economy'[6].

In this controversy of origin of agriculture Braidwood coined the term '*Natural habitat zone*'. According to him 'around 8,000 B.C. the inhabitants of the hills around the *Fertile Crescent* had come to know their habitat so well that they have been collecting and hunting'. He suggested a sub era of incipient cultivation and domestication of animals leading through final era of hunting – collecting to the primary sub era of effective village farming communities.[7]

Leslie White is one of the first scholars who have suggested an ecological approach to the origin of food production that could be tested in the field. He argues that primitive hunters and gatherers already possessed this knowledge (of food production) through their intimate association with flora of their habitat. He suggested that agriculture was simply a new kind of relationship between men or more properly women and plants[8].

All these theories indicate a change in climate or increase of population as responsible for origin of agriculture has been challenged by Hayden. He used the term '*resource stress approach*'[9] which is based entirely on man's progressive shifting towards small short maturation species as being the main contribution in the birth of agriculture.

The ecological approach to the study of the origin of food production is best seen in the article of Kent Flannery and Binford[10]. On the basis of current evidence this Flannery-Binford model suggests that food production began in South-Western Asia as an attempt by man to expand the ecological range of certain heavily exploited resources and first occurred along the margins of the Mediterranean and wood pistachio woodlands environmental zone and that it was a combination of climatic change and population increase which set the process in.

On the issue of agricultural origin, Bhattacharya has considered a number of factors like 'variability of micro-ecozones, relative abundance of wild grain in a limited territory or even the proximity of a desert have acted in combination with population increase to cause the onset of a process that leads to agriculture'[11].

The Neolithic stage of eastern and north-eastern India is very important in respect to its strategic location between various well established earliest farming settlement centres of Asia. So, Neolithic or early village farming culture of India, the most fundamental stage of human economics is a fascinating study in tracing the path of cultural development and exchange of a much significant epoch of human history that prepared the ground for civilization.

The fundamental problem regarding the reconstruction of Neolithic cultural milieu and its relationship with ethnoarchaeological data is more or less focusing towards orthodox ideology. The inconsistencies of collected data in most cases creates major problem for explanation. In the Neolithic cultural phase a transition from food gathering to food producing is observed, but our materials are not sufficient to demonstrate these transitional phase in this part of the country in great detail. It is also difficult to elaborate the gradual process of food production and the involvement of people in a very large scale.

It must be admitted that our knowledge regarding the Neolithic culture of eastern and north eastern India are not known to its entirety due to non systematic and exhaustive survey of archaeological sites in this area. Despite potential cultural yield from explorations and excavations, a somewhat distorted picture emerges largely because of lack of proper expertise as well as undertakings. Absence of fossil record especially from eastern and north eastern India is still another riddle. Paleo-botanical and paleo-zoological evidence are also not sufficient to reconstruct the floral, faunal as well as ecological history of this sub continent. Chronological horizon of eastern and north eastern India could not be determined due to insufficient number of absolute dates along with its inconsistent results. The so called Mesolithic or Chalcolithic leveled sites of Birbhanpur and Bharatpur, Pandurajar Dhibi or Tamluk in West Bengal are significant in understanding of early village farming culture of this region but could not be included as they belong to different cultural (Chalcolithic) bracket. Whatever materials are available for our study, could be able to demonstrate the elementary concept of our research methodology.

The basis of all cultural reconstruction includes precise and correctfully documentation of data, establishing its chronology, including the synthesis, reconstructing complete sequence of a culture and economy. The local area and the relationship with different archaeological exercise is mostly a part of reconstruction and that reconstruction is only possible if the database consist of artifacts, structure, floral and faunal remains and other informations related to subsistence pattern. The process of classification involves analysing all the categories of data.

The working hypotheses of the present archaeological endeavour are based on simple analogies from ethnographic data and utilize them in interpretation of archaeological facts. Thus it could be envisaged as a desperate attempt towards corroboration of known with unknown. The methodology that we have adopted here is essentially to understand our objective with its preliminary character. Therefore the archaeological data and ethnographical information forms the basis of the present study. Both archaeological and ethnographical data have been collected from relevant published accounts. Analysis of recorded data with the help of typo-technological study, relevant ethnographic works, related drawings and diagrams, the pattern of spatial context of the sites, displayed with maps are major methodologies of the present work. The photography and some other illustrations put forward here, are essentially by following the same methodological approach.

The Neolithic character of eastern and north-eastern India is developed in close contact with the neighbouring areas to the north, east and west. The brief survey of human progress including the change of climate, subsistence pattern, and technological innovations altogether help us to trace this new development. New discoveries continue to widen and deepen our knowledge about the pre-history of Indian sub continent. The study of ancient culture in eastern India considerably extended the temporal limits of pre-history and explains the origin of some integral component in the complex evolutionary process of South Asia.

It was assumed at one time that no other regions of the Old World shows evidence of any similar settled life depending on agriculture like the nuclear zone of Anatolia, Iran and Syria before 7000 B.C.E. Natural environment, suitable for food production were available in this *fertile crescent*. From recent researches in Western Asia, a fairly good number of sites have yielded evidence of early date of domestication of plant and animals. Abu Hureyra (Levant) has yielded evidence of cereal cultivation in Mesolithic (called epi-palaeolithic) in 8,500 B. C.E. and Ganj Tepe (Iran), dated to mid 8th millenium B. C.E., has yielded evidence of cultivated barley. Again at Zagross, Tepe Guran cultivated barley has been dated to 6000 B.C.E. In North-Western Asia most important site is Mehrgarh, which yielded evidence of cultivated barley and wheat during 6th millennium B.C.E. Evidence of cultivated rice has also comes from Indian site of Lahuradewa (district Sant Kabir

Nagar, Uttar Pradesh) dated to 6th millennium B.C.E. and Bajpur (Orissa) in Mesolithic context.

Vavilov in his identification of seven world centres of origin of cultivation of plant has included India and Central Asia as two separate zones amongst them[12]. T.T.Chang also identified the area of the origin of domesticated variety of rice to a broad belt stretching from central Ganga Valley to south China[13]. There is a strong archaeological evidence for early rice domestication from Yuchanyan, Hunan province and Diaotonghun, Jiangxi province, China which have been dated to 13,364 - 11,869 BP and 15,322 - 14,392 BP respectively[14].

Sauer[15] expressed that agriculture did not diffuse from one centre to the rest of the world. He identified seven centres in the Old World and four centres in the New World, where diffusion of crop plants took place. According to him South-East Asia comprising Thailand, Malaya and adjoining areas were the oldest centres of diversity and origin of cultivated plants. Besides that, forest clearance probably for agricultural purpose have been evident from Java and different parts of Sumatra. In Vietnam, rice was cultivated during early 2nd millennium B.C.E. Charcoal and grass pollen from Khok Phanom Di in Central Thailand suggest incipient agricultural activities in this region from as early as 5th millennium B.C.E[16]. All the above scholars have stressed on identifying India, as the most likely place for the cultivation of domesticated cereal and domestication of animals.

It will be worthwhile to discuss briefly about the picture of Neolithic culture and related problem conceived in the literature of Indian prehistory for getting a valid background of our own subject concerned. The summary of data available in the published works may be conceived to explore our own parameter and its relationship with human behavioural pattern existed in the ecological background of eastern India. The unevenness of development from hunting, gathering, foraging economy to pastoral agricultural economy limit us to get a coherent picture of our subject concerned. However, a brief study may outline relevant theoretical concept and its relationship with the level of our interpretative database. In the following pages, we shall try to summarise the data which might have played important role for the development of Neolithic culture, economy and the totality of a society in a given area.

The area north (Kashmir region) and west (Vindhyan Ganga region) of the study area are of special interest in our problem. In Kashmir – Swat region, from Gufkral though plant remain have yet to be systematically reported but the presence of wheat, barley and lentil have been noted from aceramic Neolithic level, dated to pre 2100 B.C.E[17]. Burzoham also has not yielded any direct evidence of agriculture but yield of some forage plant seeds which are usually occur along with wheat and barley, indirectly indicates cultivation of wheat and or barley here also. Besides that pollen analysis of Haigam lake in Kashmir valley show appearance of *plantago lanceota,* growth of *chenopodiaceae* and composite shrubs along with decline of pine forest, all suggesting towards onset of agricultural practice by Neolithic Kashmiris. Again an adjacent region of our study area, at the Vindhyas, one site namely Koldihwa requires special mention. Remains of rice from lowest level of three phase sequence *ie* Neolithic, is one of the most significant discoveries of this region. Electron microscopic studies reveal presence of both domesticated and wild variety of *Oryza* in the husk. Another site Mahagara also yielded evidence of rice remains though report does not state clearly whether it is a domesticated variety or not. As mentioned earlier, recently excavated site of Lahuradewa has also yielded domesticated rice remains from period I dated to 6th millennium B.C.E[18]. Again recent studies at Sanai Tal, Rae Bareli district suggest agricultural activity around 14000 year B.P.[19]

About the origin of rice cultivation in India, if we look to China, where earlier dates are encountered, intermediate zone between Belan valley and China has yet to reveal any evidence of rice agricultural though Gufkral in the Kashmir valley Neolithic (datable to 2800-2500 B.C.E) phase has yielded evidence of cereal farming of other cereals than rice.

In eastern India, Chirand in Bihar is the most promising site. Evidence of cultivated rice comes in the form of husk in the burnt clay, besides other plant remains. Another north Bihar site Senuwar, where carbonized rice grain from its lowest level datable to the beginning of 3rd m B.C.E. or earlier, refute extraneous origin of agriculture in India. In Bihar, from Oriup also impression of rice grain comes from Neolithic level[20]. In Orissa domesticated rice evidence has been obtained from late Neolithic phase of Baidipur. Evidence of cultivation of rice have recently been reported from aceramic Mesolithic deposit of Bajpur in Orissa.[21] The wild rice variety (*Oryza nivara*) in northern and north-eastern India has suggested being areas of understanding the origin of food production in monsoon Asia (Sauer, Vavilov). Physical evidence of domesticated pollen grain though yet to be discovered from this region, which might be due to abundance of production of wild rice that was sufficient to sustain Neolithic population of north-eastern India. Further Saraswat's comment 'even the wild looking varieties may have under cultivation and an incipient phase of cultivation, the plant remain wild in all morphological and behavioural context' (Tripathy, V. 2010 *Archaeology of the Ganga Plain* : Need for a Paradigm Shift, vol I, P 16) may be applicable here.

So, in the light of present research, it is not possible to depend on a single world theory for agricultural origin. It can be said that in disprate ecological zone all over the world, cultivation might have originated simultaneously as Binford suggested 'there were a number of independent loci of the development of cultivation technique – the Near East, Asia, the New World ..... the loci were distributed across widely different environmental types – root crop in the tropics and cereals in the semi arid lands'[22]. It seems *fertile crescent* was the land of origin of wheat and barley and East Asia was the birth place of rice.

Scholars are not unanimous even on the question of chronological priority between the domestication of plant and domestication of animal. In fact these two phenomenons are very closely related. Some authorities hold that cultivation is older than stock raising, but the German school, specially, believes that while one group started cultivating plants other group was domesticating animals. A uniformity in occurrence of different characters or traits in a given time frame or culture is not feasible. On the origin of animal domestication Burkitt wrote that animals were forced to approach nearer and nearer to man, who could then with very little effort tame them and turn them to his uses.

Among all evidences of early domestication in West, North-West, South-East Asia and India, it could be said that there is remarkable dearth of understanding early Holocene animal exploitation. In North-Western Asia, Mehrgarh, an important site which greatly helped in understanding the transition process from food collecting stage to food producing stage, but could not take us back to very beginning of pastoralism. By the late half of the 6th m. B.C.E. evidence of domesticated sheep, goat, and cattle have been unearthed. Bone of domestic goat from West Asia dated 7,800 B.C.E. has been found from Tepe Asiab. In South east Turkey at Grittile domestic goat, sheep along with cultivated plants has been dated around 7000 B.C.E. Domestication of sheeps from Chemi Shanidar, Zawi (Iran) have been dated around 9th millennium B.C.E. while domestication of goat from Ganj Dareh (Iran), Jarmo, Jericho are dated little later. From Catal Huyuk (Turkey) cattle domestication is recorded around 7th millennium B.C.E.[23] By late Neolithic period of China domestic animals (pig, dog, and ox) have been encountered from Long-Shan and other north Chinese sites.

As regards animal domestication in India Mesolithic Ganeshwar, Bagor, Adamgarh, Tilwara and Bhimbhetka are the earliest sites. At Bagor, Thomas[24] claimed the presence of domestic caprines from microlithic period I (somewhere between 5000 B.C.E. – 2800 B.C.E) to early historic time, thus suggesting an early hunting – gathering, pastoral economy in this region, though this identification was not accepted by some scholars. At Gufkral, Burzahom domestication of animals has been attested from early Neolithic layers datable to 2800 B.C.E. – 2500 B.C.E. The Neolithic folk of Utnur was largely, pastoral, who kept cattle, sheep, goat, during 3rd m. B.C.E. Adjacent to our study area, from Belan valley at Mahagara domestication of sheep and goat have been identified during 1770 B.C.E. – 1545 B.C.E. Domestic animal bones are also evident from advance Mesolithic level of Langhnaj. A very precise and cautious identification by advanced technique of bone remains of wild animals of period I (microlith) and domestic animals of period II (Chalcolithic) from Loteshwar in north Gujarat are significant. According to Meadow and Patel[25] these specimens may give some new dimension to the question of origin of animal domestication, if the previously determined date of 3700 B.C.E – 2900 B.C.E. again appear after re-examination of the charcoal sample from Chalcolithic layer.

In the present state of research it could be said that much lacuna are there in the early Holocene faunal record of this subcontinent. We can atleast authenticate that domesticated cattle (*Bos indicus*) found from Mehrgarh and India are different in species level from that reported at Anatolia and Thessaly dated to 7th millennium B.C.E. On this basis scholars suggested that this *Bos indicus* might have evolved either from local wild cattle (*Bos namadicus*) of Pleistocene period or through hybridization of imported form of *Bos taurus* with *Bos namadicus* [26]. So an indigenous origin of humped cattle in India is an established fact.

In case of sheep and goat, it is found that sites yielding earliest evidence of these two claimed domesticated animals *ie*. Bagor and Bhimbhetka do not come under wild sheep, goat breeding belt of western Himalaya. Besides that cytogenetic studies evidently show that all modern domestic forms of sheep have evolved from a single ancestor *Ovis orientalis*. According to Meadow[27] the first domesticated animal of Mehrgarh was brought from Baluchistan or Iranian Plateau, who exported those domestic animals to further east and in this way sheep and goat might have reached Indian subcontinent.

In eastern India few domesticated animal remains have been unearthed. They mostly come from Chirand besides Bharatpur of West Bengal and Gopalpur of Orissa. The so-called Neo-Chalcolithic layers of many a sites have yielded evidence of domesticated animals wich include humped cattle (*Bos indicus)*, Indian domestic buffalo (*Bubalus bubalis*), domestic sheep (*Ovis aries*), domestic pig (*Sus scrofa*) and domestic dog (*Canis familiaris*)[28] which are detailed in chapter III.

The persuasion of the data regarding the beginning of domestication of plant and animal in the present study area, it could be said that this change of economy could be very well interpreted with the help of floral and faunal study of early phase of BRW levels of different sites of eastern India. As we know that there are a sizeable data analysed from the collection of different sites like Pandurajardhipi, Bharatpur, Mahisdal, Sonepur etc which indicate towards a settled village farming economic life though we cannot incorporate them in our study as they are slightly later in date.

Production of food from animal and plant remains led to surplus which led to necessity of food storage. Childe considered this manufacturing of pottery as a universal feature of Neolithic community[29]. In Western and North West Asia, earliest plant and animal domestication centres are aceramic in character. From China and Japan the earliest pottery has been ascribed a date of 21,491 and 17,441 B.C.E. respectively. In India, at Mahagara pottery is evidenced in advanced Mesolithic level of 9th – 8th millennium B.C.E. In eastern India and north eastern India many sites yielded evidence of pottery of different types in Neolithic cultural phase. The characterstic cord-impressed ware of Vindhyan region is also found in many eastern Indian sites like Chirand, Chechar Kutubpur, Kuchai etc., though it differs in colour, in the range of decorative pattern from that of cord-impressed ware of Daojali-hading and Sarutaru.

In Indian context Paleolithic and Mesolithic are very much central, western and Deccanese phenomena while north-eastern India provides handful evidence of a Neolithic phase. Scholars all over the world have opined that archaeology of north-eastern India can play a very crucial role to the development of World archaeology, but it has largely remained unrepresented and scored almost blank in archaeological map of India. Nevertheless this region has not received due attention of scholars possibly because Sir Alexander Cunninghum, the father figure of Indian archaeology, did not concern to venture in the forest clad north-eastern India. Thus Indian archaeology which largely followed his footsteps, did not develop much interest in the regions of north eastern India, covered by the states of Assam, Meghalaya, Arunachal Pradesh, Nagaland, Manipur, Mizoram and Tripura, even few decades ago. After a phase of long negligence, the post independence period has witnessed some degree of institutional undertaking in archaeology of north-eastern India and by sixties the Neolithic background of north-eastern India was confirmed. In case of eastern India, it could again be said that as Cunningham did not take any interest in pre-history, thus eastern Indian prehistory also did not receive proper attention for years. Under this situation it is the crying need of the time to overcome the long backlog in prehistoric research work of this region and the present work is one such attempt towards this direction.

Archaeological activities in eastern and north-eastern India were initiated by the Britishers like V. Ball, E. H. C Walsh, J. Cockburn, S. E. Peal, J. H. Hutton, P.O.Bodding, Captain Beeching, John Anderson *et.al.* But sincere and incessant efforts are being made from sixities and seventies when Indian scholars like H.D.Sankalia, V. D. Krishnaswamy, P. C. Dasgupta, A. H. Dani, B. K.Thapar, D. Sen, T. C. Sharma, M.C.Goswami, and many others contributed significantly in attaining Indian prehistory a recognition.

The antiquarian wealth in India first attracted the attention of the officers of East India Company, in the last quarter of the 18th century, though at that time archaeology in the sense of the investigation and study of relics of the past formed only a negligible part of the activities. Whatever limited archaeological works were undertaken in the form of reports was not on scientific line because those officers were not trained in this discipline. Besides few valuable works, most of them were mixed with legend and touched by a sense of wonder. As early as 1842 a 'stone-knife' was discovered from Raichur area[30]. In eastern India the earliest discovery of Neolithic tool was made by E.H.C. Walsh from Kalimpong sub-division of Darjeeling district in the year 1865[31]. In the same year Capt. Beeching reported first Neolithic from Chaibasa, Chakradharpur in Singhbhum District[32]. In 1865 and 1867 V. Ball reported stone implements from Bihar[33]. In 1867, he found stone celt from Gopinathpur in district Manbhoom[34]. In 1867 John Lubbock reported a highly finished perfect celt from upper Assam[35]. In 1870 Steel described about three celts from village of the Nagas [36]. In the same publication Ball described about a celt from Buradih, south-east of Gomaria Tamar[37]. In 1875 Godwin Austen reported a celt found at Khasi hills[38]. Ball made mention about Orissan neoliths in 1876 from Dhenkanal, Ungul, Talcher and Sambalpur[39]. Polished celt, a battle axe and a scraper were reported from Hazaribagh in 1878[40]. J. Cockburn published few Neolithic specimens from Khasi hills in 1879[41]. In 1888 J. Wood Mason described about few neoliths from Ranchi District[42]. In 1889, he described about an interesting double-edged celt from Jashpur state in Chhotanagpur District[43]. During cutting through a Neolithic funeral ground Cockburn discovered few Neolithic implements from south Mirzapur District[44]. Rev. P.O. Bodding wrote about Neolithic stone implements from Santal Pargana in 1901[45]. In 1909 more neoliths were described by him[46]. In 1913 H.C. Dasgupta described shouldered implement from Assam[47]. J.C. Brown reported about grooved stone hammer from Assam in 1914[48]. In 1916 Sarat Chandra Roy described about forty-five Neolithic implements individually[49]. In 1917 C. W. Anderson's article on prehistoric stone implement also vividly described a large number of stone implements from Singhbhum[50]. In 1918 a polished hammer-stone from Singhbhum was described by H.C. Dasgupta[51]. R. D. Banerjee described about Neolithic adze from Abor country in 1924-25[52]. In the year 1926, J. H. Hutton described a celt found from Naga hills[53]. The Brahmacharis found two Neolithic celts from Jamalpur in Munger[54]. After a long silence, Orissa was archaeologically highlighted by N. K. Bose in 1940[55]. In 1948, Dharani Sen discovered some celts and chisels from a mound of village Bamal, Jhargram sub-division, Midnapur District[56]. In later publication he worked mainly on Singhbhum neoliths[57]. Y.D. Sharma reported about some chisels found from Keonjhar District, Orissa in 1952[58]. Tamluk Neolithic assemblages were discovered by M. N. Deshpande[59].M.C.Goswami and A. C. Bhagabati described about the Neolithic tools of western Assam[60] and Garo hills[61]. More Neolithic tools from Midnapur were published in the year 1962[62]. In the 64-65 season, the most famous eastern Indian Neolithic site, Chirand was first excavated[63]. Another important site of West Bengal, Susunia was brought to light by P. C. Dasgupta and was reported in 1966[64]. Besides Assam, Meghalaya, Nagaland and Arunachal, the remaining three states of north-eastern India were less highlighted. Though archaeological activity in Manipur was started as early as 1929, it suffered a long silence from archaeologists for about 32 years. From 1967[65] onwards several writings of O. K. Singh throw light upon archaeology of the state again. Under the supervision of M. C. Goswami and T. C. Sharma of the Guwahati University the Garo hill region of Meghalaya was extensively explored and reported in three consecutive issues of *IAR*[66]. A number of Neolithic sites like Citra-abri, Selbalgiri (I and II), Michimagiri, Thebrongiri were discovered during that survey· among them Selbalgiri was excavated. In 70-71 session, excavation work at Sonepur, in Bihar was resumed [67]. In following year ASI, eastern circle excavated Bharatpur in Bardhaman district and few Neolithic celts were reported[68]. The department of Archaeology and Museum, Government of Bihar conducted an exploration along the Subarnarekha valley and Neolithic celt and mace head were found from Ghoranagi[69]. Dumurkonda, Dhuliapahar in Midnapur were explored by the Calcutta University in 75-76 [70]. An important site Chechar Kutubpur in Vaishali district, Bihar was excavated by mid-eastern circle of ASI, which yielded Neolithic evidences[71]. In 81-82, Chakrabarti discovered three Neolithic sites namely Dabha, Besra and Jilintar in Purlia district[72]. In 82-83 excvation at Taradih in Gaya district was started[73]. A. A. Ashraf excavated a Neolithic site, Parsi-parlo from lower Subansiri district of Arunachal Pradesh[74]. In the same issue of *IAR* another site of Bihar *i.e.* Maner in Patna was reported. The site was excavated under the aegis of Patna University[75]. The department of Archaeology, Government of Manipur located a prehistoric site Nongpok-Keithelmanbi in Senapati district which was published in 84-85 issues of *IAR*[76]. Tripura remained almost a *terra incognita* for years. N. R. Ramesh's paper on stone tools of Tripura has thrown welcome light on the Neolithic culture of this region[77]. Excavation at Chungliyimti, Tuensang district confirmed beyond doubt that the site was inhabited during the Neolithic period[78]. In 92-93 report, a trial excavation at Pyntherlangtein, Jainta hill district which yielded a deposit of Neolithic cultural milieu, was published[79].

In all these efforts mentioned above, works mainly centred round the collection of tools, potteries and their typological analysis. As an initial move though these efforts no doubt pushed Indian archaeology few steps forward, still it has a long path to travel to attain maturity. But why have we suffered so chronically? To search for a convincing reason, Bhattacharya argued that we have always refused to incorporate the knowledge of allied disciplines in executing our archaeological problems [80]. P. Mitra once wrote in absolutely unambiguous words that 'the value of ethnography to prehistoric archaeology is very great; the two branches of science are infact complementary to each other[81].

It has already been shown that both eastern and north-eastern India are rich store houses of Neolithic materials. This region also accommodated many a tribe who still subsist on some primitive economic pursuits like hunting, gathering and other distinctive traits, unaffected more or less by modern civilized societies. So, in the present study of reconstruction of cultural life of eastern and north-eastern Indian neolithian, the knowledge of ethnography might be an appropriate aid to the fragmentary archaeological data. With this initiative the present work has been undertaken to give a comprehensive picture of Neolithic cultural life of eastern and north-eastern India (comprising the states of West Bengal, Bihar, Jharkhand Orissa, Assam, Meghalaya, Arunachal Pradesh, Manipur, Nagaland, Mizoram and Tripura) by collecting and scanning the enlisted materials and correlating them with ethnographic parallels. Efforts will be made to provide further interpretation to the existing facts and deduce reasonable conclusion about the cultural pattern of the Neolithic people of the region under study.

In doing so, the subject will be discussed in the following six chapters :

Chapter I : Introduction

Chapter II : Geographical Background

Chapter III : Archaeology

Chapter IV : Ethnography

Chapter V : Ethnoarchaeological Study

Chapter VI : Conclusion

Neolithic, the youngest of the earliest human culture, was an age of drastic alteration in cultural as well as economic achievement of mankind. Scholars were allured to work on different aspects of this crucial cultural phase to satisfy the immense curiosity about this first effort to conquer nature. So, in the Introduction chapter the criteria of the Neolithic have been reviewed in the light of eminent scholars' opinion. Along with this, we have a glance on the previous studies on Neolithic in eastern and north-eastern India, till now.

Subbarao once told, "the archaeological picture that emerges closely follows the pattern laid down by Indian geography"[82]. Again the layers of earth have preserved the fossilised past in its bed from time immemorial. So before entering into details of our main subject we will be acquainted with the geographical and geological nature of our study area and how they have influenced the pattern of culture in this region, in chapter II.

Period	Epoch	Formation Name	Rock Types
Quaternary	Recent (Unclassfied)	Newer or Low Level Alluvium	Clay, sand, silt and shingle
		......Unconformity......	
	Pleistocene (Unclassified)	Older or High Level Alluvium	Clay,coarse sand, shingle, gravel and boulder deposit
		......Unconformity......	
Tertiary	Pilocene Dihing Group	Dihing Formation	Pebble bed, soft sandy clay, conglomerate, grit and sandstone
		......Unconformity......	
	Mio-Pilocene	Dupi Tila Group Dupi Tila Formation	Sandstone, mottled clay Grit and congliomerate: beds of coal conglomerate
		Namsang Formation	and poorly consolidated sand with layers and pockets of pebbles.
		......Unconformity......	
	Tipam Group	Girujan Formation	Mottled clay, sandy shales, and subordinate, mottled, coarse to gritty sandstone
		Tipam Sandstone Formation	Blush grey to greenish coarse to gritty, false bedded, ferruginous sand stone clay, shale and conglomerate.
		Boka Bill Formation	Shale,sandy shale, silt stone, mudstone, and lenticular,coarse ferruginous sandstone.
		Bhuban Formation	Alternations of sand stone, sandy shale and thin conglomerate and shaly in the middle part

Period	Epoch		Formation Name	Rock Types
			....Unconformity....	
Tertiary	Miocene	Tipam Group	Girujan Farmation	Mottled clay sandy shales, and subordinate, mottled, coarse to grittly sandstone.
			Tipam Sandstone Formation	Blush grey to greenish coarse to gritty, false bedded, ferrugious sand stone clay, shale and conglomerate.
		Surma Group	Boka Bil Formation	Shale, sandy shale, silt stone, mudstone, and lenticular, coarse : ferruginous sandstone.
			Bhuban Formation	Alternations of sand stone, sandy shale and thin conglomerate and shaly in the middle part.
			....Unconformity....	
	Oligocene	Barail Group	Renji Formation	Massive and bedded sandstones; Its equivalent, - the Tikak Parbat Formation, in Upper Assam is marked by thick coal seams at the basal part.
			Jenum Formation	Shale, sandy shale, and carbonaceous shale with interbedded hard sand stone; its equivalent-the Baragolai Formation, in Upper Assam is marked by numerous thin coal seams.
			Liasong Formation	Well bedded, compact, flaggy sandstone and subordinate shale its equivalent-the Naogaon Formation, in Upper Assam is marked by thin bedded, hard sandstone and interbedded shale.

Period	Epoch		Formation Name		Rock Types
	Geosynclinal		Shelf		
Eocene	Disang Group (Not divided)	Splinetery dark grey shales and thin sand stone inter beds	Jaintia Group	Kopili Formation	Shale, sandstones and marl.
				Shella Formation	a) *Sylhet Sandstone Member.* Sandstone, clay and thin coal seak b) *Sylhet Limestone Member* Fossil iferous limestone.
	..........Unconformity..........				
	Pre Canbrian	Shillong group		Not classified	Quartzite, phyllite and schist.

According to Sharma 'The Neolithic industry of this area is based on maximum utilization of indurated shale, a locally available rock. Other sparsely used raw materials in this zone are sandstone and fossil wood....... Among the surface finds some elegant specimens of axes, made of jadeite are also seen[16]. The evergreen tropical forest of the area with its abundant wealth of wild edible flora as well as various fauna provided ample supply of food to present day tribals and must be stone age hunter – gatherers also.

MEGHALAYA

The state of Meghalaya was described by the British as the 'Scotland of east', The state's boundary is demarcated by the Goalpara, Kamrup districts of Assam in the north and the south western Goalpara and part of Rangpur district (Bangladesh) in the west, Mymensingh and Sylhet districts (Bangladesh) in the south and the north Cachhar and Karbi – Anglong districts of Assam in the east. The plateau stands as a watershed between the Surma valley of Bangladesh on the south and the Brahmaputra valley on the north. The rivers and streams run in straight course. The magnificent gorges scooped out by the rivers in the southern part of Khasia and Jaintia hills.

Geographically, the state could be divided into two regions – i) Eastern Meghalaya ii) Western Meghalaya. Shillong is the highest peak of eastern Meghalaya. In this region a large number of perennial streams *viz*. Umngi, Umngot and numerous small ones flow. The western Meghalaya is conspicuous with low hills and plains. Tura and Arbella are the main ranges of this region.

The state experiences tropical monsoonal climate. The world's rainiest spot Cherrapunji and Mowsynram are situated here. The agri-horticultural products of the state are rice, jute and potato. The state has abundant outcrop of very fine grained igneous dyke rocks, which could be very easily worked into various kinds of tools and weapons. Further the hill topography provides suitable climatic background for settlement.

The geological formation of the state is in following order:

Geological Age	Group Name	Formation Name	Rock Types
Recent	Newer Alluvium	unclassified	sand, silt, clay
Pleistocene	Older Alluvium	unclassified	sand, clay, pebble, gravel, boulder deposit
Mio-Pliocene	Dupi-Tlla Group	unclassified	Mottled clay, sandstone and conglomeraite.
Oligo - Miocene	Garo group	Chengapara formation	Sandsiltstone, clay marl.
		Baghmara formation	Felspathic sandstone, pebble conglomerate, clay, silt clay
		Kopili formation	Shale, sandstone, marl.
Eocene	Jaintia group	Simsang formation	Siltstone-sandstone alternation.
		Sheila formation	Sandstone, Limestone.
		Langpara formation	Calcareous shale, sandstone, limestone
Upper Cretaceous	Khasi group	Mahadek formation	Arkose,
		Bottom conglomerate formation	Conglomerate, arkose.
		Jadukata formation	Sandsotne-conglemerate alteration.

Geological Age	Group Name	Formation Name	Rock Types
Jurassic (?)	Sylhet trap	-	Basalt, alkali, basalt, rhyolite, acid tuff.
Pre Cambrian	-	Intrusive Shillong group	Pegmatite, aplite, quartz, epidiorite, basalt, dolerite. quartzite, phyllite congomerate.
Archaean	-	Gneissic complex	Biolite gneiss, migmatite, mica-schist, quartz, schist etc.

In Garo hill neolithic tool collection, it is found that intrusive igneous rock, dolerite was extensively utilized.

ARUNACHAL PRADESH

Arunachal Pradesh, the fascinating 'land of rising sun' was earlier known as North-East-Frontier-Agency. Situated along the north-east frontier of India the area has great strategic significance. The whole state is covered with eastern Himalayas and is bounded by Bhutan in the west, Tibet in the north and north-east, Mynamar in the east and south-east and Assam in the south. The entire area towers over the plains of Assam in the shape of a horse-shoe. The rugged terrian has now snow clad peaks, deep river gorges, swift flowing streams and river valley. It has a narrow belt of foot hills along the border with Assam. The entire north eastern region of India receives the heaviest rainfall in the country. The state is situated in a highly seismic zone and experiences tremors of moderate nature to violent upheavals. Its topographs, flora and fauna offer a wide variety. Most important rivers passing through this state are the Siang, Kameng, Subansiri, Dibang, Lohit, Dihing, Tirap etc.

For convenience of description the stratigraphic succession are presented in two different parts.

(A) Kameng, Subansiri and Siang Himalaya;

Recent - Sub Recent		- Alluvium and terrace deposits	
Piestocene		Ziro valley deposits; younger parts of Dihing beds (e.g Deban beds), and uppermost Siwaliks.	
Pliocene	Siwalik (Upper units)	Dihings Namsangs	
....	....	Unconformity...........................	....
Miocene	Middle and Lower Siwalik units		(equivalent to Tipams. Surmas)
....	....	Unconformity...........................	....
? Oligocene to Palaeocene		Kimi beds, sandstone and red shales, silty beds with black rock and chert fragments	(? equivalent to Barail-Disang (Dewa-thang)]
..........?..........	....	..........?..........	....
? Mesozoic		Basic trap rocks (lateritised), white quartzite	(equivalent to ? ? Sylhet Traps)
..........?..........	....	..........?..........	....
Permian	Gondwana	Sandstones, slates (with plant remains e.g. Glossopteris etc.)	equivalent to Damuda (Barakars.)
Carboniferous to Permo-Car boniferous	Bichom Group	Marine ash green shales, calc-shales etc. (With the fossils like Bryozoa, Lamelli branchs and Brachiopods). Boulder slates and Qurtzites	? Rangit pebble bed
....	....	Unconformity...........................	....
Silurian to Devonian		Miri formation - variegated slates with calc-beds.	
? Lower Carboniferous		Abor volcanics and pink and whitish quartzites.	

Lower Palaeozoic	Tenga Formation	Bands of dolomitic and carbonaceous slates; whitish quartzite and conglomeratic beds; hematitic phyllites and quartzite.	(equivalent to Buxa)
		Grey, gritty and schistose phyllites with whitish quartzite intercalations, chlorite Talc schists.	(equivalent to Daling)
..........................?	....	....?	....
Early Palaeozoic Precambrian (?)	Bomdila Group	Dirang and other schists (Kameng) Potin and Khetabari schists (Subansiri, Siang)	(equivalent to Dalings)
		Ziro, Chakoo, Sepia, Daporizo, augen banded gneisses etc. (Palaeozoic gneisses of metasomatic nature ?)	(equivalent to Darjeeiing gneisses etc.)
Palaeozoic to Precambrian	Sela Group of Higher Himalaya	Tourmaline granite (post tectonic) Teritary ? Schists, migmatites, lit par-lit gneisses etc.	Units above crystalline thrust

(B) Lohit Himalaya :

Plio Pleistocene	Upper Siwaliks	Upper Teriary rocks (Dihings)	
Mioeocene to Palaeocene	–	Not exposed	
Mesozoic	–	--do--	Covered in Lohit plains
Upper Palaeozoic	–	--do--	
Middle Palaeozoic	–	Rocks of Miri formation (Dibang valley)	

Pre-Permian Diorite-granodioorite complex = metasomatic and intrusive phase

Lower to Middle Palaeozoic	Tenga Formation? Miri Formation	Carbonates, Tezu and Tidding limestones etc. and quartizites Metasediments-(Flysch?) and serpentinites of Tidding and Tellu confluence area.

Early Palaeozoic to Pre-cambrian	Bomdila Group	Para-gneisses, high and low grade schists (Pelitic and graphitic) of Sewak Pass, Lalpani etc.

A number of neolithic sites of Arunachal Pradesh have been located in river valleys of the Kamla, Lati, Tally, Dhing, Noa dihing etc. In Arunachal Pradesh, the only excavated site Parsi-Parlo has revealed utilization of local raw material for manufacture of most of the Neolithic tools. About 61% tools were made up of locally available sandstone, basalt, gneiss, schist, talcose and quartzite. Rests of the tools were made from non-local jadeite rock.

MIZORAM

Mizoram occupies a very strategic position in the north-eastern corner of India, having an international boundary of about six fifty miles with Myanmar and Bangladesh. Geographically, the state lies in the eastern most part of India bordered by Bangladesh and Tripura in the west, Myanmar in the east and south, Manipur and Cachhar district of Assam in the north. Mizoram has the most hilly terrain in the eastern part of India which is geologically very young. The hills are steep and are separated by rivers. There are innumerable streams, rivers and brooks in the territory. The area falls under the direct influence of monsoon. Average heights of the hills are about 900 metres. There are few small plain areas in Mizoram, which are used entirely for rice cultivation. The Tuirial, Tuivawl, Mat, Tian, Kau are important rivers. The tropic of cancer runs through Mizoram therefore the climate of the state is warm and humid. Vegetation growth is abundant here. Rice is the most important crop besides maize, pulses, sugarcane etc. The forest houses various wild animals.

The general geology is represented by a repetitive succession of neogene, arenaceous and argillaceous sediments. The general lithosratigraphic succession is represented by the Barail group, Surma group, Bhuban formation and Bokabil formation. Soil type generally found in the state is a clayey loam mixed with shales. Archaeology of Mizoram is not mentionable and so also the distribution of sites and raw material of tools.

MANIPUR

Manipur is a isolated hilly terrain having distinct entity in geographical characteristics. It is a small state of beautiful hills and dales in the north-eastern border of India. Lying between Assam, Cachhar, Myanmar and Chittagong, she covers a small area of about 0.7% of the total area of India. Manipur can be divided into two geographical regions, the valley and the surrounding hills. The geographical structure of the state is of oval shape and slightly slanting towards the south. Manipur is also regarded as a rich store house of natural beauty. Highest peak of Manipur hill area is the mountain Tenipu (2,995 metres). Two types of river flowing over this area. One flowing towards the east falling into Chindwin and the other flowing towards west into the Barak river. The Manipur valley lying in the central part of the state is formed by the alluvial deposit and is drained by the Manipur river and its tributaries.

Neolithic culture of Manipur is found both in river valleys and hills of the state. The universal prevalence of dense and impervious forest has restricted our knowledge of the geology of the country. Geologically it can be called as a virgin land. General lithographic sequence of the state is Cretaceous limestone, Disang, Barails, Surmas and Tipams.

Geological Age	Group Name	Rock Types
	Tipam	Arenaceous rock
Miocene	Surma	Agrillaceous rock
Upper Eocene and Oligocene	Barails	Abundance of carbonaceous material
Lower and middle		Greyshale, mudstone, silt stone,
Ecocene to upper	Disang	sandstone buff coloured shale,
Cretaceous		siliceous limestone, grit, conglomerate.

In Neolithic industry of Manipur sandstone, quartzite sandstone, chert, serpentinite were mostly used.

NAGALAND

The state of Nagaland is encircled by Assam in north and west, by Myanmar and Arunachal Pradesh in the east and Manipur in the south. It forms an irregular plateau with elevated ridges and peaks. In consonance with the trend of ranges of spur, the courses of stream are also generally aligned in a north-south direction. The rivers have cut the hill ranges at suitable points almost at right angles, resulting in barbed patterns. The central portion commands a view of the open rolling mountains but the eastern and south western portion have a more complicated terrain. The north-western portion is a plain territory. Nagaland has a typical monsoon climate with variants ranging from tropical to temperate conditions. Numerous hill streams flow over the state. Principal river in the east is the Tizu, Dzulu in central part, Dhansiri, Diphu, the Disangfojam etc. in the west.

In Nagaland also, Neolithic inhabitants preferred the river valleys for their habitation.

Tertiary sequence of the state has been classified in the following way[20].

Geological Age	Group Name	Rock Types
Pleistocene	Alluvium	Clay, coarse sand , gravel
Pliocene	Dihing series	Loose sandstone, clay, pebble beds.
Mio-Pllocene	Dupitila (Namsang beds)	Soft coarse sand stone, clays occassional coal conglomerate.
Miocene	Tipam series	Mottled clay, fine grained sandstone.
	Surma series	Sandstones, carbonaceous shale, coal seams.
Oligocene	Barail series	Sandstone, shales, flaggy, bluish grey hard sandstone.
Eocene	Disang series	Grey splintery shales and salt

In Nagaland neolithic industry, serpentine rock and jadeite rocks predominate.

TRIPURA

Geographically, the state of Tripura lies in the eastern part of India bordered by Bangladesh to the west, south and north, by Assam to the north-east and by Mizoram to the east. The landmass of Tripura merged from the sea in the Tertiary period. Topographically the state is characterized by hill slopes, hillocks, flat lands, rivers and lakes. The terrain is mostly undulating and more than one third of total area is covered by forest. Geographically, Tripura is divisible into 2 broad regions i) Tripura plains and ii) Tripura hills.

In Tripura plains, flat lands are found mostly towards west and south. Besides that some forest clad ridges and valleys are also found. Major river of this region are the Khowai, Haora, Gomati etc. The Tripura hill region is composed of high, low and level plains. Three major hills ranges are i) Jampui ii) Longtharai and iii) Altharamura ranges. The Juri, Deo, Khowai are important rivers which made the river basin fertile to grow rice, jute and other vegetables. Whatever few stone age sites in Tripura are discovered, all are located on the elevated ground on the river banks.

The general geology of Tripura reflects, teetonically a part of mobile trough. The rock formation of Tripura range in age from lower Tertiary to Recent.

General geological succession of the state is —

Group	Subgroup	Formation	Rock Types
Recent	—	Recent	Alluvium represented by silt, sand, clay, etc.
Dupi Tila	—	Dupi Tila	Sandy Clay, sand loam
Tipam	Tipam Sandstone	Champanagar	Sandstone, sandy-shale
		Manu Bazar	Sandstone, silt stone, mud stone.
Surma	—	Bokabil	Sandstone, shale, mud-stone.
	—	Bhuban	Sandstone, shale, siltstone.

About 98% of the entire collection of Tripura stone tools is made of silificied wood, which is chiefly procured from Tipam sandstone formation.

So, from each and every material relic among the few survivals of the neolithic culture it is apparent that the primitive men had sufficient knowledge about their surrounding environment and they used it most effectively. It is a fact that man never lived suspended in mid air, he laid his feet solidly on this earth from which he procured his food and raw material. It could be said that early man was a successful practising ecologist.

In the study of early human culture, early environmental contribution could not be denied. With the acquired knowledge of extant vegetation we can reconstruct the vegetational history and palaeo-environment of Quaternary period. Unfortunately, in India the palaeoenvironmental study is not very significant yet, except Kashmir Rajasthan and the Ganga valley. If we accept that change of climate was not a local phenomenon, we can discuss the studies conducted so far in India.

From pollen profile it has been suggested that in 17000 B.P., Kashmir valley climate was hot while from circa 10000 B.P. to the present, the valley enjoyed a cold temperate to hot temperate season[22]. In circa 10000 to 9500 B.P. the climate of Rajasthan experienced a high rainfall and wet climate. From circa 9500 to 5000 B.P. the area received a low rainfall season and was followed by a highly humid climate in between circa 5000 and 3000 B.P. From 3000 B.P. onwards environment was not unlike the present day *i.e.* dry to sub dry. In Thar desert the begining of Holocene was moist for 500 years then upto mid it was warm with increased rainfall.

In West Bengal the only excavated mesolithic site, Birbhanpur, has thrown some light about the contempory climatic condition. On the basis of mechanical analysis of fine particles, microscopic examination of sand grain and chemical analysis of the relevant soil samples, it has been suggested that the pre microlithic and non-implenetiferous layers of the site 'were under conditions of large quantities of moisture and high mean annual rainfall with a downward percolation of water'[24]. The microlith bearing layer marked a comperatively dry period with increased wind activity. The lower Gangetic plain has so far yielded little indication of its climatic past. During Holocene-Pleistocene boundary two dominant arid phases and weak monsoon were identified. But from the uniform high clay size component from Holocene section of eastern Bay of Bengal, it could be inferred that during Holocene, strong monsoon with a humid climate prevailed. From a number of pollen sample study in and around Calcutta, it is suggested that around 5000 years B.P. there was a sporadic agricultural practice in the man made swamp land, besides that pollen grains of *plantago* also suggest presence of human settlement in that period around Calcutta[25]. The claim of geologists about the presence of human being around Calcutta in neolithic period has not yet been proved archaeologically; still its importance should not be ignored. In the Darjeeling Himalaya, marked expansion of broad leaved forests indicates the onset of a period of optimal climate which continued until 4000 years B.P. After 4000 B.P. the monsoon climate began to weaken in South Asia. Winter precipitation also seems to have been reduced and mean annual temperature fell.

The palaeo environment of eastern and north-eastern India could not be reconstructed acurately, due to lack of systematic excavation. It is generally accepted that after the cold and dry phase of the terminal Pleistocene the climate was almost the same as the present day. The presence of large sized grass pollen grains along with some cultivation indicator suggests the existence of primitive agriculture in upper Assam also[26]. Detailed geomorphelogical study[27] in the Rongram, the Ganol and the Simsangiri stream area suggest that these streams have cut deep valleys in later period and have preserved two distinct fill terraces. The top surface of the older fill forms a distinct terrace in the area and a good number of neolithic sites have come from this terrace. The yellowish flood silts of younger fill suggest that these streams aggraded in response to late Quaternary drier climatic condition. The thick flood silt and and gravels possibly indicate that forest cover in the area was thin during the late Pleistocene drier climatic condition. In Holoccne, the intensity of monsoon was stronger over the north-eastern India. Increased colluvial sedimentation during the terminal Pleistocene in Garo hill area is suggestive of a drier condition than that of today.

Thus the above analysis of physiographic as well as geological history of our study area exhibits enough possibility to trace the sequence of human behaviourial pattern since early prehistoric time. At the same time the above mentioned enormous sources of human life also display the possibility of human growth including the changing pattern of human behaviour including the relationship between man and natural resources. It is obvious that evidence about the exploitation of natural resources is very meager for documentation, still it could be said that Holocene period played an extremely important role in human history. The archaeological sources which have been studied in the succeeding chapters must help us to conceive somehow the character of neolithic culture of eastern and north-eastern India.

References:

1. Subbarao, B. 1958, *Personality of India: Pre and Proto Historic Foundation of India and Pakistan,* Baroda, P.8
2. Spate, O.H.K. 1967, *India and Pakistan, A General Regional Geography,* London, P.522
3. Banerjee, B, 1972, 'Morphological Regions of West Bengal', *GRI,* XXVI (IV), P. 14.
4. Walsh, E.H.C. 1904, 'A Note on Stone Implements in Darjeeling District', *JASB,* LXXIII(III), pp. 20-24.
5. Geology and Mineral Resources of West Bengal, *G.S.I Publication* no. XXX, (I), 1974, P.3
6. Datta, A. 1992, *Neolithic Culture of West Bengal,* Delhi, pp. 67-119.
7. Geology and Mineral Resources of West Bengal, *G.S.I, Publication* no XXX(I), 1974, P.76
8. Hunday, A and S. Banerjee, 1967, 'Geology and Mineral Resources of West Bengal *MGSI,* LXXXXVII, P. 302.
9. Ahmed, H. 1965, *Bihar : A Physical, Economic and Regional Geography*, Ranchi P. 14
10. Geology and Mineral Resources of Bihar, *G.S.I, publication* no. XXX, 1974, pt V, P.5
11. Sinha, B.N. 1971, *Geography of Orissa*, N. Delhi, P.5
12. Geology and Mineral Resources of Orissa, *G.S.I. Publication* no. XXX (III), 1974.
13 Dash, R.N. 2000, 'The Neolithic Culture of Orissa : A Typo- Technological Analysis' in K.K. Basa, P. Mohanty (eds), *Archaeology of Orissa,* Delhi, pp.201-202.
14. Pithawala, M.B. 1942, 'Physiographic Divisions of India, *SC* VII, pp. 533-543
15. Geology and Mineral Resources of Assam, *G.S.I publication* no XXX, (IV), 1974, pp.35-38.
16. Sharma, T.C. 1981, The Neolithic Pattern of North Eastern India in M.S. Nagarajarao(ed) *Madhu : Recent Researches in Indian Archaeology and Art History,* Delhi, P.42.
17. Oldham, T. 1984, *Geology, Metrology and Ethnology of Meghalaya*. P.22
18. Geology and Mineral Resources of Arunachal Pradesh, *G.S.I Publication* no. XXX(IV), 1974, P. 10
19. Brown, R. 1975, *Statistical Account of Manipur,* Delhi, P.4
20. Bareh, H. 1970, *Nagaland District Gazeteer,* P.6
21. Geology and Mineral Resources of Tripura, *G.S.I, publication* no XXX, (IV), 1974, pp. 117-120.
22. Singh, G. 1990, 'Environmental Changes In South Asia During Holocene'. *JP* XVI, P.277.
23. Rajguru, S.N. 1983, 'Problems of Late Pleistocene Aridity in India', *ME,* VIII, pp. 127-134
24. Lal, B.B. 1958, 'Birbhanpur : A Microlithic Site in the Damodar Valley', West Bengal, *AI,* XIV, pp. 4-48

    – Chakrabarti, Dilip.K. 1999, *India : An Archaeological History,* N. Delhi, P.95.
25. Chanda, S and B.B Mukherjee, 1982, 'Radio Carbon Dating of Two Microfossiliferous Quaternary Deposit in and Around Calcutta', *SC,* XXXV, P. 275.
26. Bhattacharya, K.and S. Chanda, 1987-88, 'Late Quaternary Vegetational History, Palaeoecology of Some Deposit of Brahmaputra Basin, Upper Assam : India', *JP,* XXIII-XXIV, pp. 225-237.
27. Rajguru, S.N. 1981, Geoclimatic Background of Prehistoric Assam' in V.S. Srivastava(ed) *Cultural Contour of India*, pp. 186-198.

Chapter 3

Archaeology

The antinquarian wealth of India started attracting the attention of East India Company's officers from the middle of the 18th century[1]. In 1842, a Neolithic tool in the form of a 'stone knife' was discovered by Primrose[2] from Lingsigur area of Raichur district of Mysore. Following that in 1850 about twelve Neolithic celts were discovered from Uttar Pradesh by H.P.Le Mesuriar[3] and gradually series of prehistoric sites were discovered in southern, central and western India. But at the very initial stage, those limited archaeological works did not provide every detail and thus Neolithic character of those artifacts could not be properly established. Wheeler's excavation at Brahmagiri[4] in northern Mysore in 1947 revealed a sequence of culture of which period-I is now designated as Neolithic or Neolithic–Chalcolithic cultural period in the south Indian context.

So far as eastern India is concerned, due to some Christian dilettantes, this part of the country was brought to our notice by the late 19th century and first Stone Age implement, a Neolith was discovered from Kalimpong sub-division of Darjeeling district of West Bengal in 1865[5]. Later discoveries from north-eastern India proved the area as mine of stone artifacts, especially in Neoliths.

This chapter has a brief discussion on general features of sites related to our specific study, not only in terms of exclusive evidence but also to get a clear idea about the distribution pattern of Neolithic and sub Neolithic sites. We have very meager evidence of the existence of settlements of agricultural community in our study area. The distribution pattern of sites showing the movement of Neolithic groups is by and large very tentative altogether. However, it is significant to note that all these sites carry useful material to understand different aspects of neolithism including the incipient stage of agriculture, the beginning phase of well defined pastoralism, the refinement of hunting, gathering, foraging, fishing and finally the survival pattern of Neolithic folk in later period. In this context we may note that most of the Neolithic sites provided us multifaceted data to understand the basic character of Neolithic and the role of agro-pastoral economy in our study area. The occurrences of stray neoliths which have been recorded as Neolithic sites sometimes confuse us about their character. But it is our assumption that all these sites played important role for the day to day manoeuvering of Neolithic folk in eastern and north-eastern India.

Krishnaswamy initially classified four neolithic provinces in India but with more discoveries now scholars have broadly rccognised six different Neolithic zones in India *i.e.* 1) Northern region (covering Kashmir valley), 2) Belan valley region (covering Vindhyan plateau), 3) Mid-eastern region (covering north Bihar), 4) North-eastern region (covering Assam, Meghalaya, Manipur, Mizoram, Tripura Arunachal, Nagaland and the sub-Himalayan region), 5) Central eastern region (covering Chhotanagpur plateau with its peneplains extending in West Bengal and Orissa), 6) Southern region (covering the peninsular India). To begin with, the regional Neolithic pattern of the sub-continent is surveyed in brief in section-I followed by an account of further details of the present study zone *i.e.* mid-eastern and north-eastern and central-eastern regions in section-II

Section -I

(1) *NORTHERN REGION* - Northern Neolithic culture is represented by two type sites (Burzoham and Gufkral) besides others. This northern Neolithic shows certain unique characteristics which single it out from other Neolithic cultures of India.

Burzoham reveals three phases within the Neolithic occupational period. Period-IA is aceramic in character. Pit dwelling is the most distinctive trait of this phase. Circular and square pits were used for habitational purpose. Apart from these unique structural remains the site has also yielded bone and stone tools. Both the types are prolific and include arrow head, harpoon, needle, point etc. in bone and axe, adze chisel, hoe, harvester, drill, point etc. in stone. Archaeological evidence of wild species of animals *viz.* ibex, bear,sheep, goat, cattle, wolf, stag etc. as well as of domesticated animals *viz.* sheep and goat have been encountered. The culture shows considerable emphasis on hunting. Plant remains include wheat, barley, lentil, peas etc. Period-IB is an evolved stage of earlier phase, without any break. No significant change in life pattern is noticeable except the introduction of hand made pottery. Neolithic potters of this area made graceful pots though the use of wheel was not known to them. Potteries are of three types-thick coarse grey ware, fine grey ware and gritty dull red ware. In Period-IC mud and mud brick structures replaced the earlier pit dwellings. A new wheel made ceramic variety with black burnished surface made its appearance. Varieties and better finished bone and stone tools appeared in this phase, though typologically these were a continuation of period-IA. Few painted pots were discovered towards the close of this period. Alongwith these, a few copper made objects have also been found. From the upper level of this period the discovery of about 950 beads of carnelian and agate seems to be of separate context. Traces of inhumation and secondary burial were found in this phase. Besides that a remarkable practice was found in the provision of regular burial of animals. Other finding includes engraved stone slabs depicting hunting scenes. ^{14}C date for the north Indian Neolithic culture indicate a time bracket of 2400-1500 B.C.E (uncalibrated)

(2) *BELAN-VALLEY REGION*- The Belan valley region along with its adjacent Vindhya and the Ganga valley has now been established as a potential prehistoric settlement of India.Notable Neolithic sites of this region are Koldihwa, Mahagara Chopani-mando, and recently discovered Jhusi, and Lahuradewa.

Koldihwa Period I belongs to Neolithic period. Stone tool industry of this region is represented by celt, ringstone and microlith. Ceramic industry comprises of hand made, cord impressed ware, rusticated and burnished ware. Husk of domesticated rice grain has been found in the paste of pottery clay. One of the ^{14}C dates from the Neolithic deposits of Koldihwa is 5440 $\pm$ 125 B.C.E. Mahagara is a single culture Neolithic site. Here a number of huts, arranged in nucleated fashion, have been found. In the hut floors were found microlith, Neolithic blade, celt, bone arrow head, quern, muller, terracotta beads and bones of animals. Pottery assemblage constituted of hand made, cord impressed, rusticated burnished red and burnished black wares. Notable discovery of the site is a number of hoof impressions of cattle of different age groups within a fenced area, suggesting the site as a cattle-pen. In Chopani Mando, a proto Neolithic phase is represented by tranchet, hammer stone, anvil, quern, ring stone etc. Two types of hand made pottery *i.e.* red ware and brownish grey ware, sometimes decorated with impressed designs are evident here. Besides, a number of hut foundations and hearths were noteworthy discovery of this phase. Excavation revealed remains of wild rice and bones of wild cattle, goat/sheep. The entire occupational deosit of Chopani-mando have tentatively been placed within 17,000-7,000 B.C.E. Further north at Jhusi in Allahabad district and Lahuradewa in Sant Kabir Nagar district, Uttar Pradesh have yielded a Neolithic culture comprised of hand made ceramic industry, beads of semi precious stone, bone tools, circular hut floor, domesticated and wild cereals as well as domesticated and wild animals etc. Importance of Lahuradewa lies in providing the earliest date (6442 B.C.E) of domesticated rice in Indian sub continent. Earliest ^{14}C date of Jhusi is 7447 B.C.E.

SOUTHERN REGION – Neolithic culture of this region has attracted the greatest attention of the archaeologists. A good number of Neolithic sites *viz.* Brahmagiri, Sanganakallu, Maski, Piklihal, Utnur, Tekkalakota, T.Narsipur, Kodekal etc. have come to light. Besides minor variations, these sites represent fairly homogeneous character.

These Neolithic occupations are mostly divisible into two sub phases. The lower sub phase is characterised by a ground industry, microlithic tool, distinct burial practice and hand made pottery. In phase II wheel made pottery appeared and construction of mud houses occurred along with the earlier traits of phase I. The common variety of tools include axe, adze, scraper, wedge, chisel, hammerstone, sling stone, quern, pick, borer etc. In South Indian Neolithic, a lithic blade industry is always found with ground tools. At Palavoy and Sangnakallu, a crude microlithic industry with large number of patinated flakes precede the Neolithic industry. As regards ceramic industry of phase I, differential firing resulted in a variety of colours- pale grey, buff and brown. Post firing red ochre coloured paintings is evident in some cases. Perforation, applique method and limited range of motif-paintings were used for decoration. In phase-II coarse brown and black ware appeared besides red ware, having painting in black. Presence of metal objects in a number of south Indian Neolithic sites and multiple urns suggest cultural contact with the Deccan Chalcolithic culture. Other miscellaneous crafts of this region include working in bone and manufacutre of beads. Besides this the neolthic culture of southern region is beset with another phenomenon *i.e.*-occurrence of ash mounds. In ash mounds heaps of vitrified ash of different sizes occurs in the form of mounds. They were probably burnings of cowdung on a large scale at high temperature to funerary holocaust. Few Neolithic and megalithic potteries are found in the ashy layers of these mounds. Main distribution of such mounds has been located in the districts of Karnataka and Andhrapradesh.

Radio carbon dates of south Indian Neolithic indicate a wide range of 2400 B.C.E. to 650 B.C.E and somewhat later if Brahmagiri dating has to be taken into consideration.

Section II

In this section all the Neolithic sites of eastern and north eastern Indian states and West Bengal, undivided Bihar, Orissa, Assam, Meghalaya, Arunachal Pradesh, Nagaland, Manipur, Mizoram and Tripura will be enlisted, with the Neolithic materials discovered and published. In this context, we may note that due to inadequacy of funds, the number of excavated sites are not very remarkable in India. Therefore our attention towards explored sites is obviously a purposeful exercise. In Bihar, Chirand, Checharkutubpur, Senuwar, Taradih and Maner, in Orissa, Kuchai and Golbai Sasan, in Assam, Daojalihading, Sarutaru in Meghalaya, Selbalgiri, in Arunachal Pradesh, Parsi-Parlo and in Manipur, Napachik are few excavated Neolithic sites (Plate 33).

EXCAVATED SITES

BIHAR

Chirand (District- Saran)

Chirand is situated on the river bank of the Ghagra and the Ganga. A large area of Chirand was excavated to get a full fledged picture of prehistoric period. The Neolithic find of Chirand Period - I include ground celts, scrapers, chisels, hammers, pestles, querns along with bone and antler needle, pin, styli, arrow head, unfinished shovel, broken comb, bangles, toy cart wheel, perforated terracotta disc etc.(Plate 10, 11, 12). Ceramic industry comprised mainly of handmade pottery though some was made of turntable method (Plate 14A, 14B; Fig.1). Some burnt clay bore paddy husk impression (Plate 13). Circular houses of wattle daub were

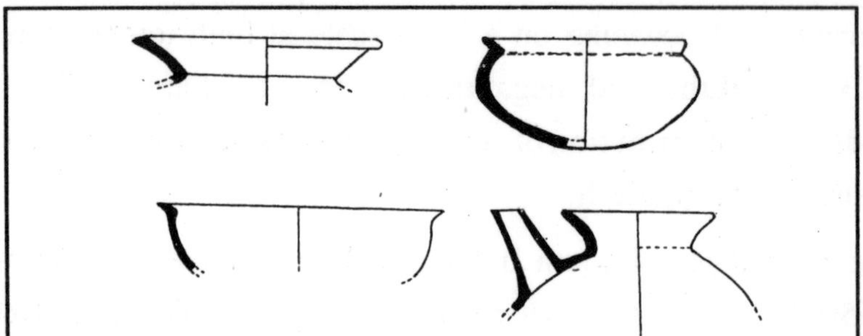

Fig.1. Neolithic Pottery from Chirand.

also among the discovery. In late phase of the culture, a circular floor and a series of open hearths have been found. The whole settlement was encircled by a 75 cm thick mud wall construction. Period - II is full blown Chalcolith *i.e.* NBP ware, two Neolithic celts and chisel, corroded punch marked coin and cast copper coin and from Period- III five phases of structural remains have been unearthed. Besides that a large structure of residential type, terracotta human and animal figure and amulet depicting Surya were also found. Period IV reflects the beginning of Christian era. Pottery of blackish grey ware characterize this period. From Period V glazed ware, red and black ware, terracotta plumb bob, glass bangle, stone pounder, and bone - disc and reused bricks have been found[6]. Available ^{14}C dates of Neolithic phase of Chirand range from 3338 B.C.E. to 1300 B.C.E.

Senuwar (District-Rohtas)

Most important Neolithic site of Rohtas district is Senuwar, on right bank of the river Kudra. Cultural deposit of Senuwar (F.2A) period IA is pure Neolithic in character while period IB is represented by Neolithic tools in association with metal. Stone tools assemblage of Period-IA comprised of microlithic blades, flakes, bladelet, drill, wide range of bone tools etc (Plate 8). Burnt clay pieces with reed marks and charred animal bone object of bivale shale form special cultural traits of Senuwar. Potteries are mostly wheel turned, while few are hand made(Plate 9). Most interesting discovery of the site is the recovery of pollen evidence of rice, barley, lentil and wheat. In continuation of Period IA, IB has yielded all the traits of Period IA *viz.* Neolithic polished celts, microlithic blades, saddle quern, rubber, muller etc. with an addition of metal. A remarkable introduction was of black basalt sharpener, besides bone point, borer, copper wires, fish hooks, needles, terracotta figurines, beads of semi precious stone etc. Ceramic industry is not much different from that of period IA. Another significant discovery of the site is a well rammed earthen floor. The beginning of the Neolithic at Senuwar has been put around 2300 B.C.E. Period II of Senuwar is assigned to Chalcolithic period while Period III is characterized by introduction of iron and NBP ware. Period IV belongs to Kushana period[7].

Chcchar Kutubpur (District-Vaishali)

A small scale excavation was conducted on the northern bank of the river Ganga at Chechar - Kutubpur. A three fold cultural sequence has come to light from excavation at Chechar. Among them period IA is comparable to Neolithic Chirand. Lilhic industry consists of polished stone tool, bone and antler tools including double forked pick axe with socket for hafting (Plate 15). Ceramic industry is dominated by red and grey wares. A significant feature of this ware is the existence of applique designs on the shoulder. Ochre painting, though restricted in rim portion, is found throughout. Partial exposure of house plan indicates that some houses were circular in plan with earthen flooring. Period IB is distinguished by the tradition of ceramics and house pattern of previous phase along with some ordinary bone implement. Period 1C has yielded evidence of pottery of red ware, black ware with shiny slip and microbead. Painting on pottery is also evident. Period II belongs to NBP phase. From period III large brick structure is encountered. Such structure first started during the Kushan period and continued upto the Gupta period[8].

Taradih (District- Gaya)

Easy availability of raw material from Rajgir hill area and Paimar valley might have facilitated the flourishing of a Neolithic settlement around Gaya district. The site of Taradih is located on the south west of the Mahabodhi temple at Bodhgaya. Excavation revealed seven different periods in its strata[9]. Neolithic assemblages of the occupational strata of period I include celt, microlithic blade, quern, pestle, lunate, cores, hammers, charred animal bone with hewn marks, bone arrow head (Plate 16A) point, disc, terracotta ball, archaic animal figurine, fish hook, arrow head, point, chisel etc. Hand made potteries offer a variety of colours and shapes. Post firing ochre painting, incised designs, and thumb and rope impression, irregular scratches decorate the pots. Ovens of different shapes were also among the discovery of this period. A chunk of burnt clay with reed impression indicates house of wattle and daub construction. Period - II of Taradih has yielded iron, NBP ware, punch-marked coin etc. with Neolithic celts of different sizes. The following period III yielded evidence of iron while period IV belongs to NBPW phase. PeriodV yielded Kushana period occupation. Period VI belongs to Gupta and post Gupta phase while from period VII, a Pala period monastic complex has been found.

Maner (District - Patna)

From Patna one important excavated site, Maner, and few other explored sites with little Neolithic yield have come to ight. Occupational strata of period I of

Maner belong to the Neolithic age. This layer has yielded microliths besides (Plate 16 B) bone points, terracotta spindle whorl (Plate 17) and a large variety of pots. Red and grey ware, cream and grey ware, were the main ceramic of this period. During the mid level *handi*, dish with short stem, stand of bowl, short necked jars, and many other varieties occurred. Period II of Maner represents Chalcolithic cultural material while Period III and IV are NBP phase and Pala in characters respectively[10].

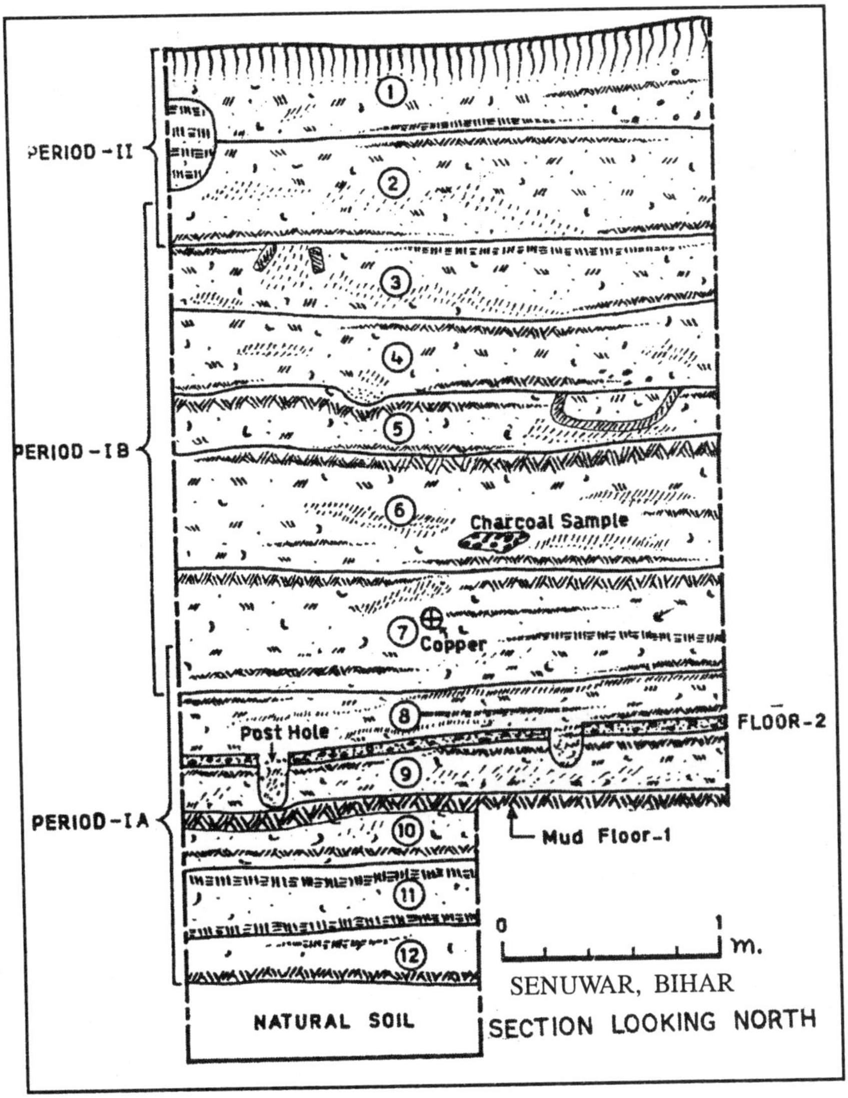

Fig. 2 (A) : Section of Senuwar, Bihar

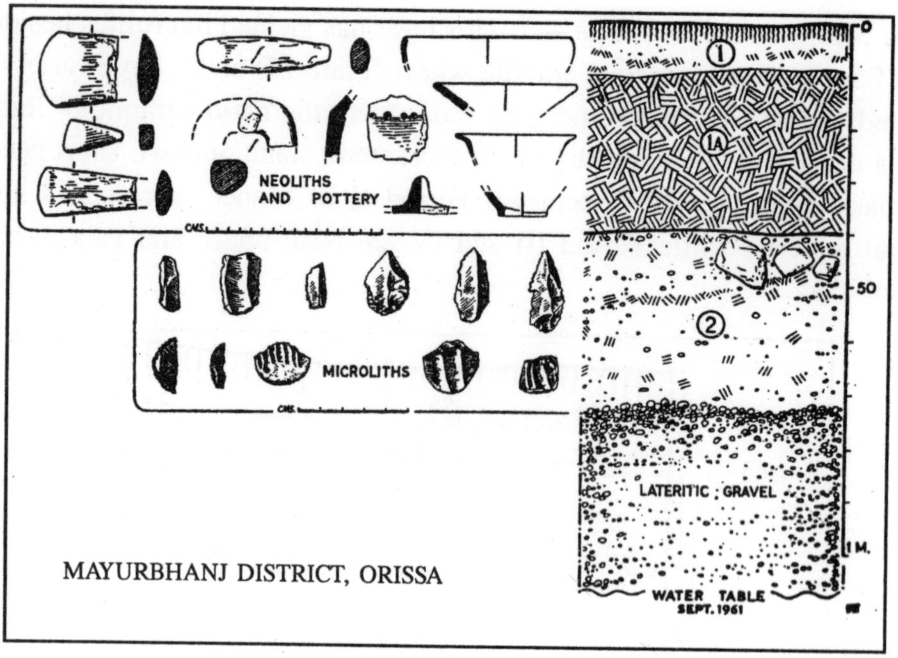

Fig. 2 (B) : Section of Kuchai, Orissa

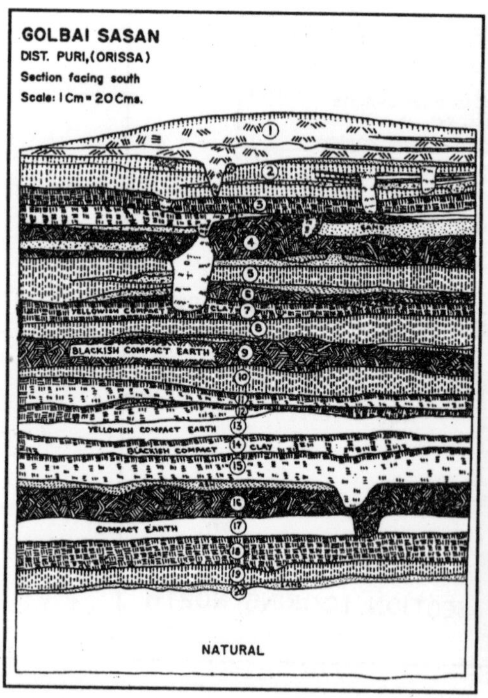

Fig.2 (C) : Section of Golbai Sasan, Orissa

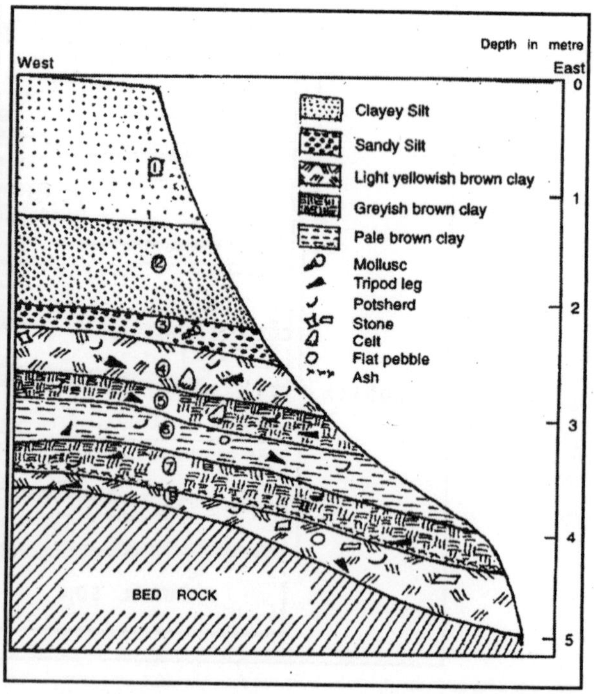

Fig. 2 (D) : Section of Napachik, Manipur

ORISSA

Kuchai (District - Mayurbhanj)

Kuchai, situated in the north of the district town of Baripada, near Burhabalang River. It was first excavated by B.K. Thapar. Excavation revealed a 40 – 45 cm clayey deposit (Fig. 2B) containing Neolithic artifacts. They were chisel, pounders, facetted hoe, mace head and grinding stone, ground stone axe of pointed end variety (Fig. 3F-3I). In association with these artifacts were found a coarse grit tempered red ware (Fig. 4A-4F) which is recorded for the first time in this region and is of very special interest of the site. Some potteries are sometimes slipped and show incised or finger tip decoration. Shouldered adze was recovered from surface. The discovery of a non-geometric microlithic industry of late Stone Age is unrecognized hitherto in Mayurbhanj district precedes the Neolithic phase.

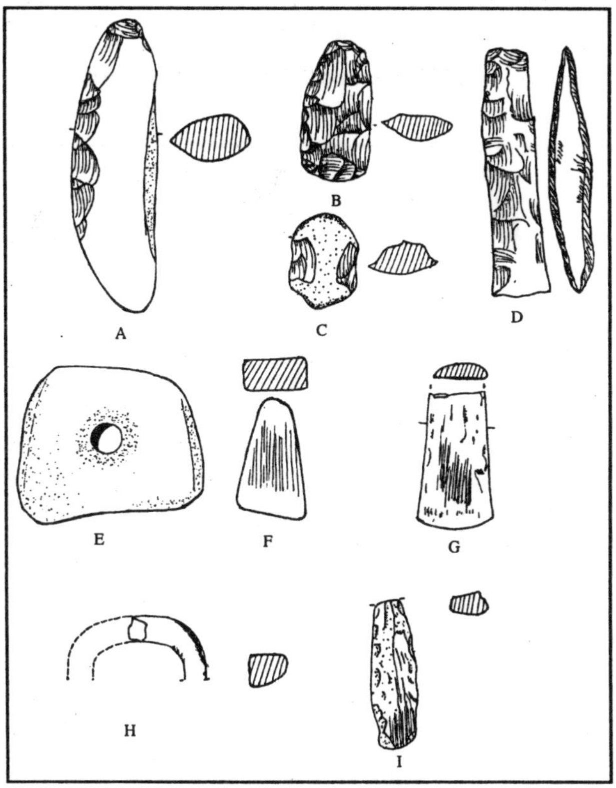

Fig. 3. Neolithic stone tools from Orissa.

A, C- Pebble tool, Bonaigarh; B- Axe,- Bonaigarh; D- Chisel, Thakurani; E-Ring stone, Gopalpur; F-G– Celt, Kuchai; I– Chisel-kuchai

A thermoluminescent date of 1000 B.C.E. for a pottery of Neolithic level at Kuchai has been reported[11].

Golbai Sasan (District - Puri)

Excavation on the left bank of the Mandakini, in district of Puri, Golbai Sasan yielded Neolithic occupation. The identified 20 habitatiional layers (Fig. 2C) have been divided into 3 periods among which Period I is characterized by Neolithic assemblage. In this period no stone or bone tools were met with, although a few pieces of bone with working marks were found. Along with this some practically mineralized bones and antlers were found. Pottery of period – I is hand made on slow wheel or turn table, with cord and reed impression (Fig. 4G-4I). Coarse dull red ware and grey ware potteries came in different shapes and sizes. Moreover a structure of unshaped stone on a bed of clay belonging to this period was also noticed in this period[12]. The Neolithic deposit has been dated to *circa* 1600 B.C.E. Period II A of Golbai has been assigned to Chalcolithic cultural period and Period II B is almost similar in character of Period II A with one iron implement.

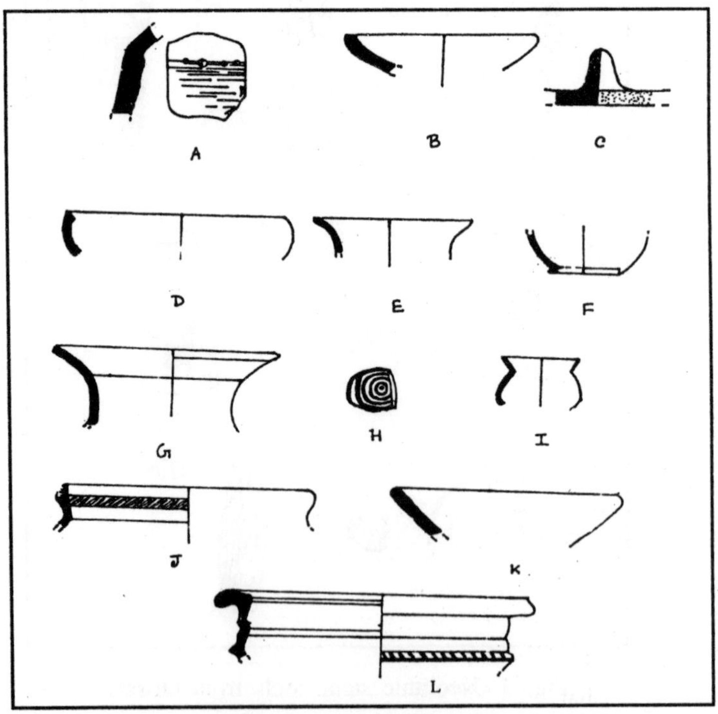

Fig. 4. Neolithic Pottery from Orissa
A-F – Kuchai; G-I – Golbai Sasan; J-L – Gopalpur;

ASSAM

Daojali Hading (District – North Cahhar hill)

The most promising Neolithic site so far discovered in north eastern India is Daojali Hading belonging to this district. It is located on the ridge of a Tertiary range in the north Cachhar hills. Excavation revealed stratified evidence of a single cultural level *i.e.* Neolithic. Lithic industry of Daojali Hading comprises of small celts, quadrangular axes, adzes, grinding stones, querns, mealing trough, muller, fossil wood etc (Plate 21B). Potteries are predominantly hand made. Grey or dull red pottery shows cord marked decoration (Plate 22). The minor types are stamped dull red variety and brick red variety. Besides these a host of household utensils were also found among the discovery [14].

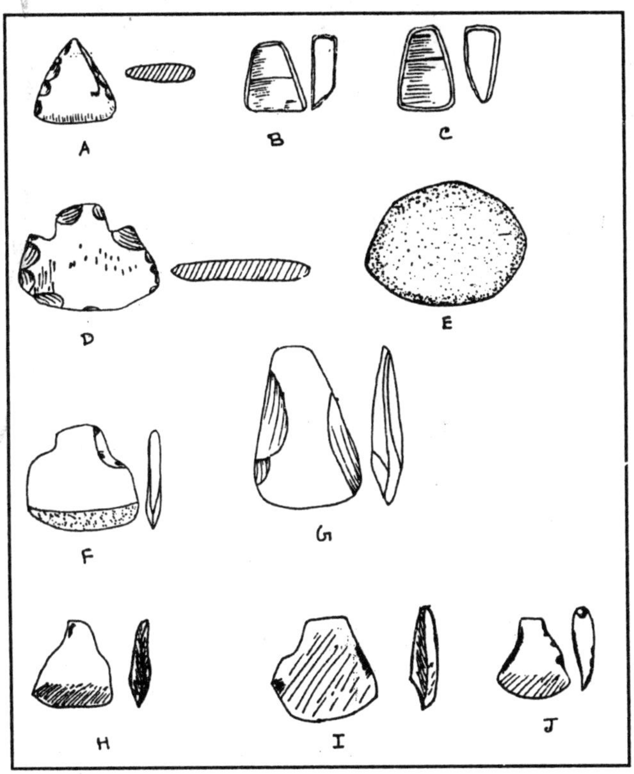

Fig. 5. Neolithic stone tools from Assam, Meghalaya.

A. Smal Celt, North Cachhar hill; B. Quadrangular adze, North Cachhar hill; C. Chisel - North Cachhar hill; D. Shouldered celt - North Cachhar hill; E. Grinding stone - North Cachhar hill; F. Shouldered Celt - Sarutaru; G. Celt - Sarutaru H-J. Shouldered celt - Garo hill

Sarutaru (District - Kamrup)

Only important Neolithic site of Kamrup district is Sarutaru. The site itself lies on the top of a small hillock flanked by a tributary of the Brahmaputra. Excavation at Sarutaru has revealed a single cultural deposit of about 35 cm. depth. Stone tools industry of the site is represented by nine ground stone celts of slate (Fig 5F, 5G), of them seven are of shouldered variety while rest are of rounded butt axe type. Fragmentary pieces of potsherds are the representative of Sarutaru ceramic industry. Pottery is mainly cord impressed and handmade (Fig-6). Brown ware predominate the industry while buff and grey ware, were also available here. Design repertory consists of parallel incised cross lines. A date of beginning of the Christian era has been suggested by the excavator[13].

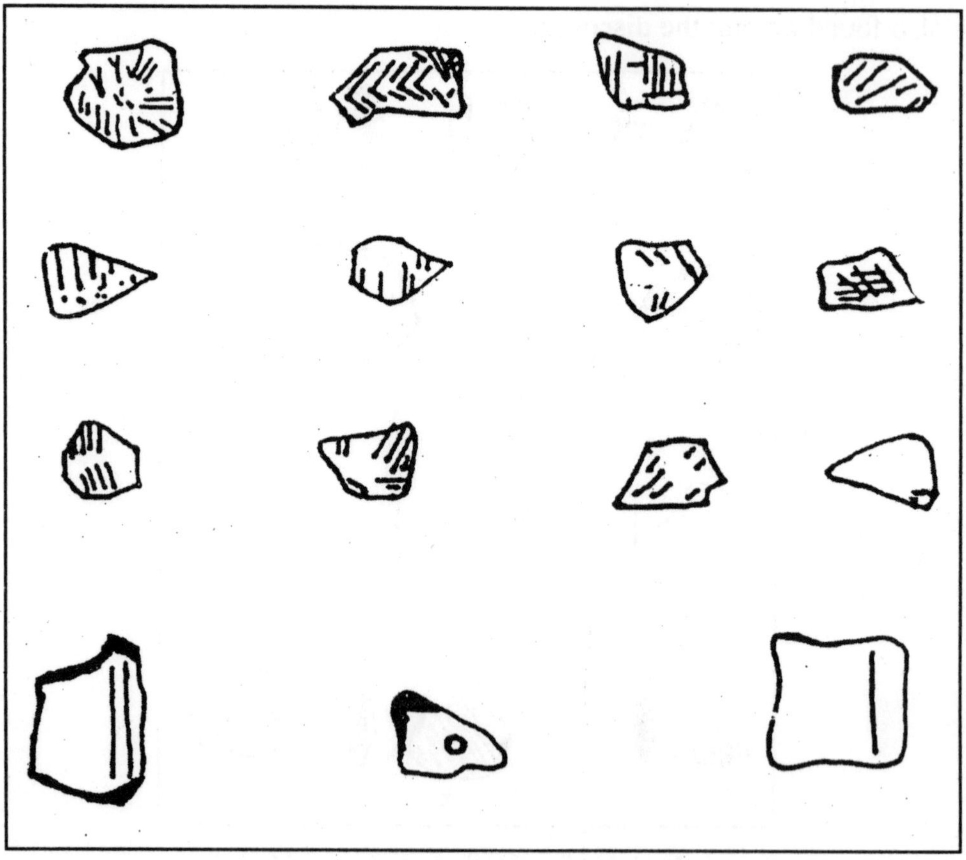

Fig. 6. Neolithic Potsherds from Sarutaru.

MEGHALAYA

Selbalgiri (Disrict-Garo Hill)

Selbalgiri is situated on the southern bank of the river Rongram. Excavation revealed a cultural deposit of 77 cm thickness. Layer I has yielded ground chipped axe, scraper (Plate 23) and sherds while layer - II has yielded evidence of a micro blade industry of flake, core, hammer and pottery[15].

ARUNACHAL PRADESH

Parsi Parlo (District-Subansiri)

Parsi Parlo is situaled on the right bank of the river Kamla. Excavation revealed about 50 - 100 cm thick cultural deposit divisible into following phases. Phase-I – aceramic Neolithic, Phase-II – ceramic Neolithic stage and Phase-III – ferrolithic stage. PhaseI is characterized by scrapers and large cutting tools. Tool assemblage of this phase is essentially Palaeolithic in character but ground and pecking technique has already taken its birth. The layer-II belongs to phase-II. This phase is characterized by pecked and ground implements waisted blade/axe, facetted tool (Fig. 12). Tools mostly possess hafting facilities. A few number of fire places with ash and charcoal have also been encountered in this phase. Hand made potteries represent honey comb, square grid and beaten impressed decoration[16].

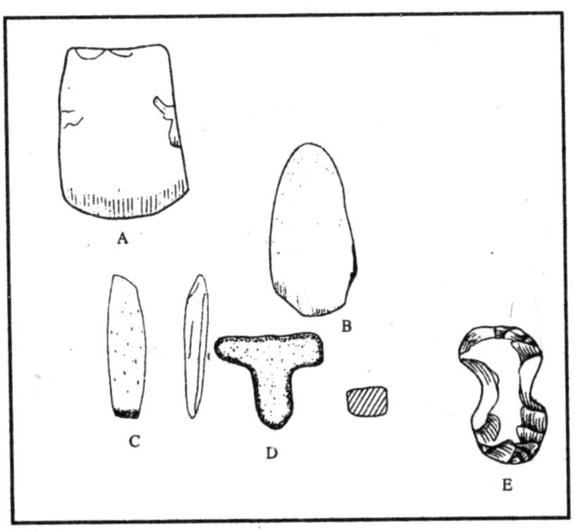

Fig.7. Neolithic stone tools from Arunachal Pradesh.

MANIPUR

Napachik (District-Wangu)

In Manipur, only one site has been excavated. On the right bank of the Turel Achouba (Manipur river), at Napachik seven layers (Fig. 2D) have been exposed[17]. Among them layer-V is represented by a small pebble chopper, a flake, a broken piece of ground tool and tripod ware pottery with cord marked decoration. Layer- VI is the thickest amongst them all. It contains a broken piece of grinding stone, an edge ground knife, flaked flat pebble and flakes and pottery of previous types (Plate 25,26). A potsherd sample from the site was dated by TL dating method. It gives an age of 1450 B.C.E. for the tripod culture in Manipur.

Besides the above mentioned sites, most of the Neolithic artifacts are found from explorations. Though exploration can not tell about the exact stratigraphic position or cultural succession of an object but in India, where explored sites outnumber excavated sites, to judge the potentiality of a site the importance of explored sites can not be ignored. Explored sites, again, can be divided into two types *viz.* one yielding sizeable amount of relics and other yielding stray finds. In India, as most of the archaeological discoveries are exploratory in nature, the antiquities bear no stratigraphical reference and it is very difficult to determine the age of the finds. Moreover, although it is found that all cultures are characterized by some typical tool types of their own, there may be cases of continuation of tool types of one period to the succeeding cultural phase because of their utility in the ecological context of a particular region. Under this situation all those sites yielding characteristic Neolithic objects have been described in this section, though all of them might not be Neolithic in date. We shall now discuss about those explored sites within the scope of our study.

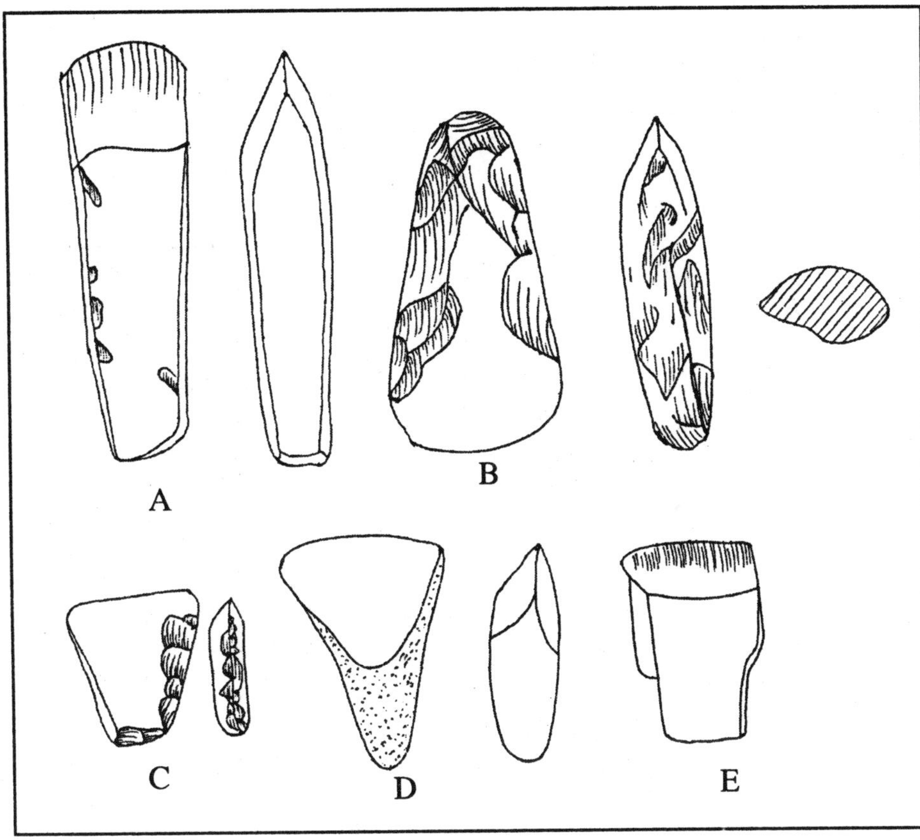

Fig. 8. Neolithic stone tools from Manipur

EXPLORED SITES

WEST BENGAL

Sankalia in his classification of Neolithic and Chalcolithic cultures of India considered Bengal Neolithic as 'pure Neolithic'[18] though Neolithic in stratified context is yet to be found. Neolithic relics are found from three distinct geographical areas *i.e* northern mountainous region, south-western plateau fringe area and alluvial tract of southern Bengal (Map 1)

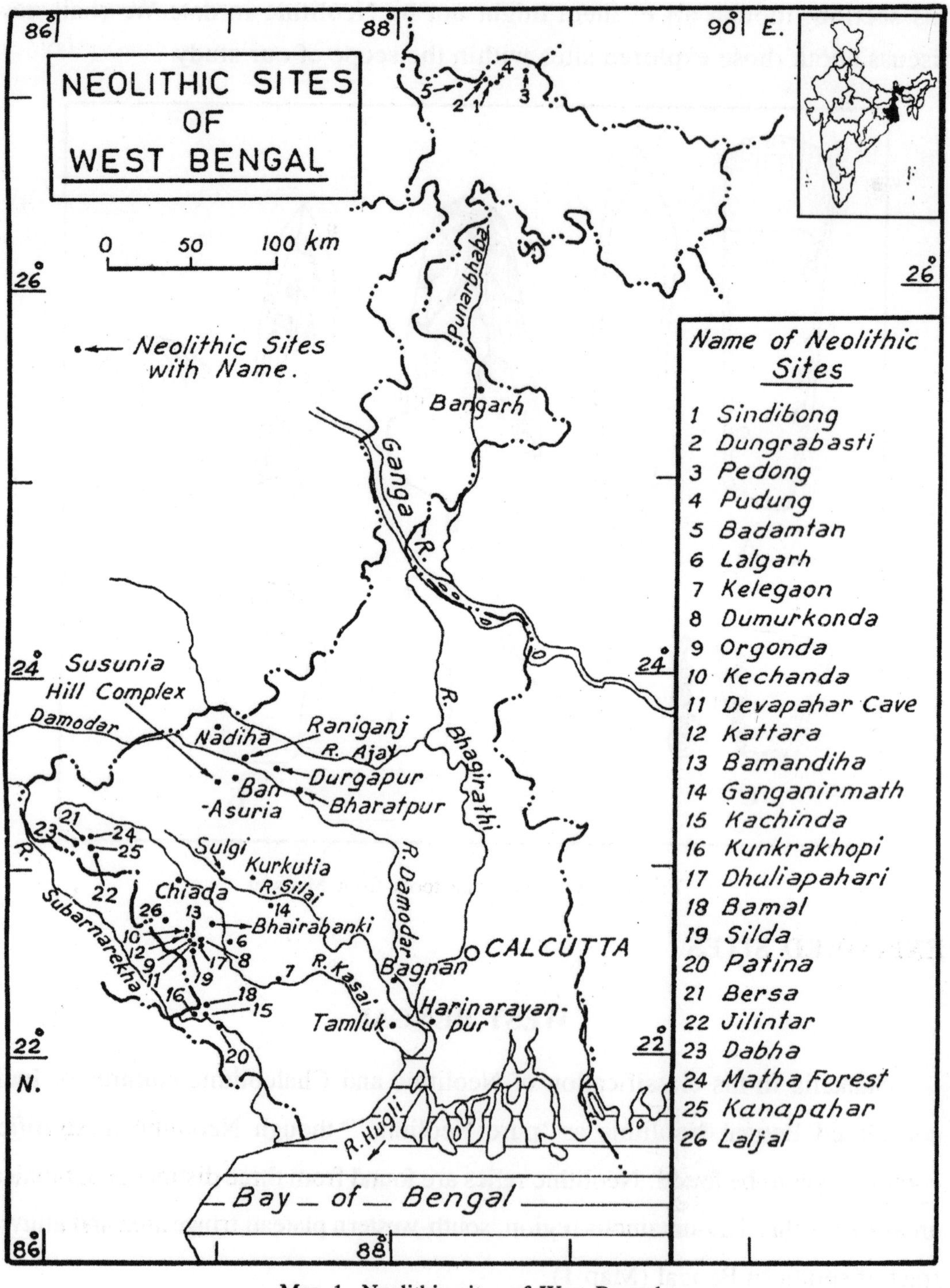

Map 1. Neolithic sites of West Bengal

Northern Mountainous region, an extension of Assam and Shillong Plateau yielded a good assemblage of neoliths. The district of Darjeeling mainly comes under this division.

Darjeeling:

In Darjeeling district stone implements are most frequently found in the Kalimpong sub division (Plates 1,2, Fig.9). Earliest discovery of neoliths was made in this region in the year 1865[19].

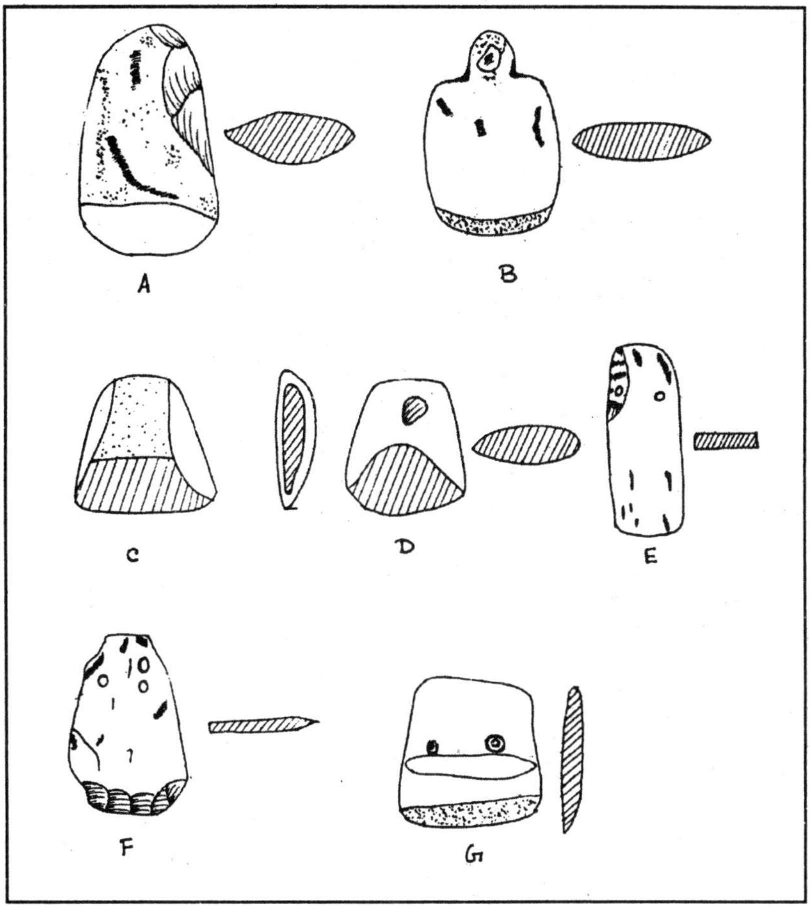

Fig.9. Neolithic stone tools from Northernn mountainous region, West Bengal A. Celt, B. Socketted Celt, C. Facetted tool, D. Adze, E.F. Chisel, G. Perforated celt.

Two polished tools *viz*. one celt, one hammer, one chisel and one adze have been recovered from east of the Kalimpong valley, at Sindibong and Dungrabusti. Some ornaments of greenish jadeite similar to that of Burzahom finds are also mentionable among the discovered objects[20].

Directorate of Archaeology, Government of West Bengal conducted extensive survey in this locality afterwards. As a result, Neolithic celt and adze with perforation were recovered from Pedong and Pudung[21] area. A chipped Neolithic industry analogous to that of South East Asia was noticed at Badamtan[22] site, afterwards.

West Dinajpur:

Bangarh is an important site, situated on the bank of the river Punarbhaba. Here excavation has revealed cultural remains of different phases ranging from 2nd-1st century B.C.E. to the Pala age. In this historical site, discovery of a rounded butt Neolithic celt in a layer just below the Sunga level (2nd-1st c.B.C.E.) is note worthy[23].

Plateau fringe area:

The area adjacent to the Chhotangpur plateau is the richest and most important Neolithic zone of West Bengal. The districts of Midnapur, Bankura, Birbhum, Purulia and Burdwan coming within this zone have yielded Neolithic evidence (Plate 3,4,5 Fig- 10)

Midnapur:

The main river system of the district is maintained by the Kangsavati, Silavati, Subarnarekha and their tributaries.

In the Kangsavati river valley, two main Neolithic sites are Lalgarh and Kalegaon. Lalgarh is an important site situated about 5km south of Bamal. Sen discovered few elongated Neolithic celts from this region. From a trial digging twelve stone implements including ground stone celts and chisels were found. One chisel and one celt were also found *in-situ*[24]. Kalegaon is situated on the eroded cliff overlooking the Kangsavati River. The site has yielded Acheulian, middle Palaeolithic and Mesolithic tools besides Neoliths. Neolithic variety includes smooth pointed butt Neolithic celt[25].

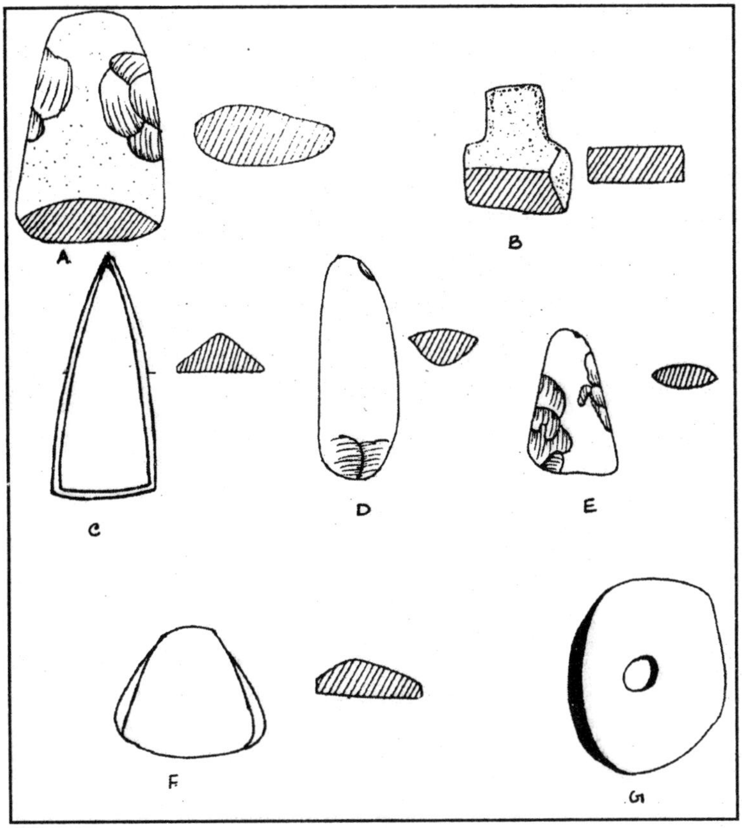

Fig. 10. Neolithic stone tools from south-western plateau fringe area, West Bengal.

A. Celt; B. Socketted celt; C. Facetted tool D. Adze; E,F. Chisel, G. Peforated celt.

In river valley of the Tarafeni, a tributery of the Kangsavati, archaeologists have come across a number of Neolithic sites like Bamandiha, Dumurkonda, Orgonda, Kechanda, Laljal, Bhulabheda etc. Dumurkonda[26] has yielded microlithic blade, triangle, arrowhead, scraper, Neolithic celt and ringstone. From Orgonda[27] microliths and neoliths have been discovered. An excavation at Kechanda[28] has revealed two Neolithic celts. A trial digging inside Laljal cave revealed one half broken ring stone, few wheel turned grey potsherds, microliths and some bone tools.[29] One shouldered celt have been found from Bhulabheda near Belpahari.[30]

From an exploration in the upland Bengal twenty two celts, sixteen ring stones; few grooved stones have been collected from the middle reaches of the Tarafeni valley.[31]

From the Silavati river valley, Ganganir math[32], on the ancient terrace near Garbeta, has been located. In the capping of red soil early, middle and late Stone Age tools were found embedded. Neolithic find of this site is a celt. Another Silavati river valley Neolithic tool yielding site is Bagridihi[33]. Four finely smoothed bar-celts have been found from this site.

From Suvarnarekha river valley, a number of Neolithic sites have been discovered. They are Murasole, Kachinda, Thakurani Pahar, Kankrajhor and Sadhanpara. Post graduate students of Archaeology department, Calcutta University, have discovered a trapezoidal celt from Murasole[34]. From Kachinda[35] two trapezoidal celts have been collected. In the year 1983-84, the Archaeology department of West Bengal Government, discovered a fragmentary Neolithic celt, a few microliths, two broken ring stones and Neolithic pottery from Kankrajhor [36], a site flanked by the river Kharosti. Neolithic yield of Sadhanpara[37] is also a trapezoidal celt.

On the bank of river Dulung, a tributary of the Suvarnarekha, Neolithic specimens were picked up from Kurakhopi[38]. From surface collection at Thakuranipahar[39] two broken shouldered celts have been reported. Besides Tamluk, Ichhapur and Amritberia are two other sites of the Rupnarayan valley which have yielded hoe and chisel[40].

Other sites of Midnapur district are Dhuliapahar, Kattara, Deulbarh, Silda, Bamal etc. At Dhuliapahar[41], 6 km. north-west of Belahadi, Neolithic cultural elements like celts, ring stones, adzes and sherds are repesented with all the Stone Age cultural representatives. Kattara[42], a valley between two hills, has yielded some broken Neolithic tools and ring stones. State Archaeology department, West Bengal recovered small Neolithic ground celts from a mound near village Bamal[43] in the Jhargram sub division. From Deulbarh, Neolithic shouldered celts have been recovered by Mukherjee[44]. Neolithic industry of Silda[45] is represented by polished stone axe and pestle. From Perua[46], a number of shouldered celts have been found.

A.Ghosh described[47] some surface collection of Neoliths from Midnapur but no particular site name has been given. Finds include a pounder, a ring stone, and two celts. A Neolithic rock shelter was discovered in Jangal Mahal area of Patina Village[48] Jhargram.

Bankura:

Neolithic sites have been reported from the river valleys of the Silavati, the Kangsavati and the Damodar and the Susunia hill region in the Gandheswari river valley area.

From the Damodar river valley, at Bon-Asuria[49] several Neolithic bar celts and few stone celts have been recovered. The Kangsavati River has provided a number of Neolithic sites in its valley. From Chiada [50] three Neolithic objects have been recovered from exploration. At Kurkutia[51] two tiny triangular celts and few sherds were found. Mentionable observation is made on a celt which was reworked after use. The sites at Kankradara[52] and its adjacent areas have yielded a developed Acheulian industry besides Neoliths. Neolithic celt, pounder, ring stone are picked up as surface finds.

From river valley of the Silavati, only one site, Shulgi, is located. Besides microliths, two Neolithic celts, few pieces of iron implements have been discovered[53].

Department of Archaeology Government of West Bengal had brought to light one of the largest Stone Age factory site in Eastern India in Susunia hill region. A small excavation at the Bhaluksonda cave, Susunia, in the Gandheswari and the Dwarakeswar river valley, reveal cultural assemblage from Neolithic to iron using culture. Along with the artifacts of bone and antler, lithic tools particularly burin, hand drill, blade and scrapers have also been found. Bone implements include harpoon, spear head etc[40]. Besides that a large number of Neolithic sites in Susunia[55] hill terrace and banks have yielded early, middle, late and new Stone Age tools. Only Neolithic ring stone have been found from sites namely Amjore, Biribari, Dhankora, Gidhuria, Hapania, Paharbedia, Paharghata and Parkul. Among these sites from Parkul, celts and from Dhankora, quadrangular axe have been encountered. Rounded butt smooth axe have been reported from Chandra. Bishinda has been mentioned for yield of triangular smooth small axe. A small Neolithic celt of white agate with red and purple vein is a mentionable find of this hill area.

Neoliths have also been found from different sites on the bank of the river Bhairavbanki[56], *viz.* Thakurbari, Fakusma, Sagarchak, Jemradanga, Patgara, Raspal, Ladsol, Jugladanga etc. Finds include mostly axe and microlith (Fig. 11).Jugladanga has yielded few ringstones, mullers, adzes, bolus stone etc.

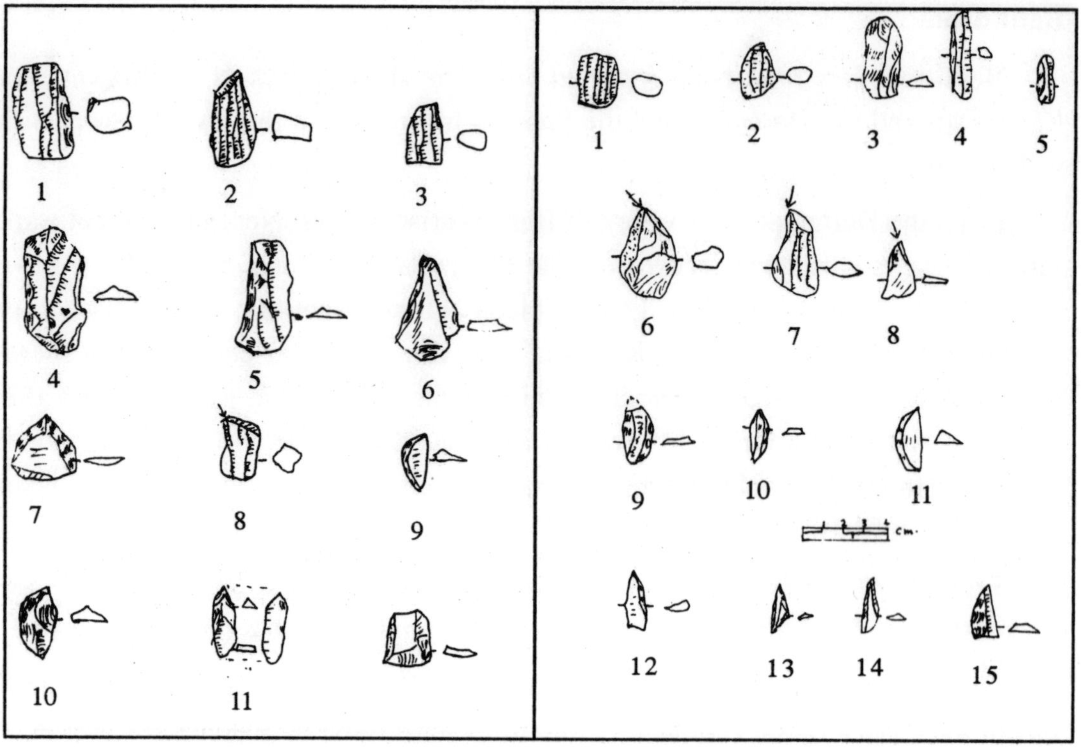

Fig. 11 (a-b). Neolithic associated microlith from West Bengal.

Purulia:

Neolithic tools have been reported from different parts of this district.

The Suvarnaekha river valley, Purulia district has yielded shouldered celts, adzes, triangular celts of Neolithic type. Besides that fossilised bone were also recovered from kankar deposit[57].

From a survey of the Department of History, Delhi University, a number of Neolithic sites like Besra, Jilintar, Dabha have come to light[58]. Besra neoliths include triangular axes with well polished working edge. Few microliths are also found along with Neoliths. Jilintar is a forest village near Ayodhya hill. One chipped and pecked triangular celt, one chipped and pecked pointed butt celt and three broken celts are the Neolithic yield of this site. Dabha is situated about 10km from Balarampur, on Balarampur Bagh-Mundi road. The Neolithic assemblage consists of triangular axe, rectangular adze, muller and pointed butt celt.

From Kanapahar[59] neoliths have been found with Palaeoliths and Microliths. From Matha forest[60] region, half portion of a ringstone has been reported. Ringstone have also been recovered from Amgora, Sargo, Hoaroka. Tikartanr Chiputnr[61] is mentioned as a Neolithic site of Purulia district. Bhattacharya[62] described two celts found near the Ayodhya hill. The Directorate of Archaeology, West Bengal found shouldered celts, adzes, triangular celts from an exploration carried out in this district[63].

Bardhaman:

Lying between the Ajay, the Damodar and the Hooghly river Bardhaman district was also a seat of Neolithic people, though in comparison to Midnapur and Bankura district, the number of Neolithic yield is lesser[64].

From Durgapur also some Neolithic artifacts were collected[65]. An exploration in Raniganj area revealed some Neolithic evidence[66]. Jakherdanga, on the confluence of the river Brahmani has yielded Neolithic tools along with post-Neolithic industry[67]. Department of Archaeology, West Bengal recovered seven Neolithic celts from different levels of Panduk[68].

Birbhum:

Very few Neolithic sites have been reported from this district. On the river bank of Bakreswar and Mayurakshi at Potanda some Neolithic celts of pointed butt end variety are found[69] with microliths, black and red ware and other characteristic pottery. Some Neolithic implements have been discovered from Jhinaipur – Keoda[70].

Parisar[71] is an explored site of Birbhum district -which has yielded Neolithic artifacts. Chakrabarti has reported[72] about a small, polished celt from the river bank of Kopai.

Murshidabad:

From Murshidabad district only a few sites with Neolithic remains have been reported.

Gitagram is an explored site. Neolithic finds of the site comprised of celts, along with Chalcolithic pottery, microlithic flakes and beads of semiprecious stones[73]. In 1966-67, two polished stone celts have been retrieved from an exploration at Rajbadidanga[74].

Alluvial Tract:

It is found that in southern part of West Bengal the Neolithic sites are mostly concentrated in south of the flood plain of the Ganges. Stray occurrence of neoliths beyond this area could hardly be called Neolithic.

From the Gangetic delta, at Nadia district one Neolithic tool has been collected[75] but detailed report is lacking. One traingular celt was recovered near Bagnan[76] railway station. An exploration by the department of Archaeology in 60-61 has brought to light pointed butt polished celt, early pottery and a large number of other antiquities from Harinarayanpur[77] of south 24 Parganas district. The village of Deulpota, near Diamond Harbour is reported to have yielded one celt, one hammerstone, two pounders and three pebbles[78]. Chandraketugarh of Barasat sub division is an excavated site though no evidence of Neolithic culture is found from digging but some Neolithic tools have been collected from surface[79].

So, it can be deduced from the above discussion that main Neolithic sites of West Bengal lie in the south-western part of the river Ganges in the alluvium and in the gneissic and lateritic terrain of extended Chhotanagpur plateau area. In northern mountainous area, the Neolithic community possibly developed separately. The Neolithic industry of this mountainous portion was devoid of any ceramic trait. Most of the Neolithic sites of this region were developed along the confluence of the rivers and streams.

So far as the alluvial tract is concerned, only a few tools have been reported, often in association with early historical finds. Though Tamluk finds include a number of Neoliths with Chalcolithic cultural material, still no mentionable Neolithic site and large number of Chalcolithic sites in the neighbouring area leads to suggest that until some Neolithic finds in stratified context are discovered it can only be said that Neolithic men did not found West Bengal suitable for permanent settlement, in south west plateau fringe area, adjacent to raw material resources area of Chhotanagpur plateau, neolithians were thinly scatterd. This alluvial tract was permanently inhabited from Chalcolithic period onwards. If the Neolithic status of Laljal cave is established then on the basis of two evidence Laljal and Jangalmahal from Midnapurs it could be said that Neolithic folk of this area atleast, occupied rock shelters also for habitational purpose.

BIHAR AND JHARKHAND

Neoliths of Bihar come from a diversified physiographical settings like extremely rocky, forested and isolated Chhotanagpur plateau, alluvial Gangetic plain and rocky hills of Kaimur region. Among them in Chhotanagpur plateau, Singbhum district and northern Bihar Jharkhand are the most promising regions of stone age culture of eastern India (Map 2).

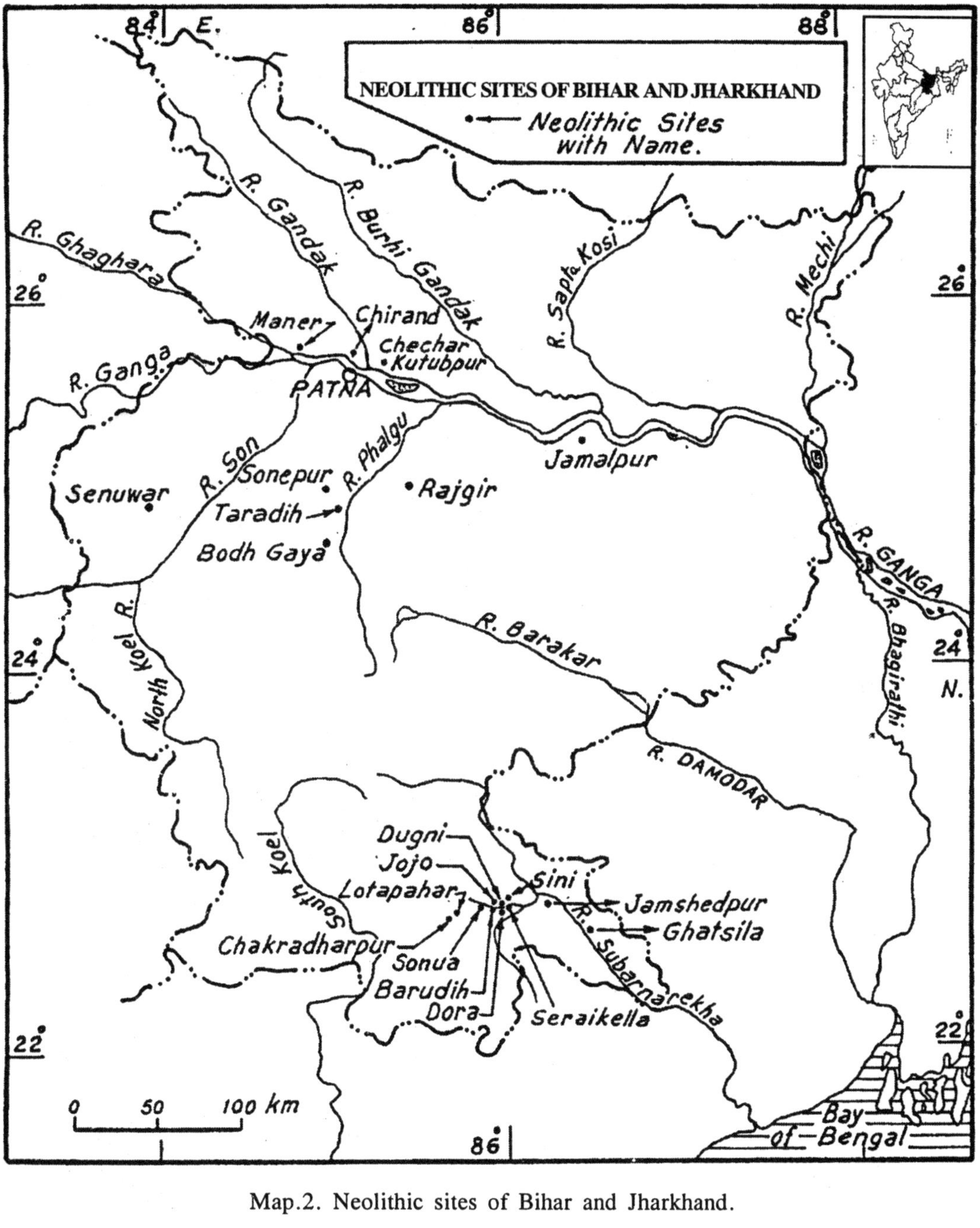

Map.2. Neolithic sites of Bihar and Jharkhand.

Singbhum:

In Singbhum district Neolithic sites are scattered over the river valleys of the Sanjai, Subarnarekha and Kharkai. (Fig. 12, Plate 6)

Trial excavation at Sonua on the right bank of the river Sanjai has brought to light. Palaeolithic, Microlithic as well as Neolithic assemblage of tools[80]. Neolithic stone industry is represented here by celts of varied shapes, fair amount of chisels, number of hammerstones, fabricators, rubbers, pounders, ring stones, fragmantary mace heads and saddle querns. About 2,500 potsherds have been collected from the present site. They are both hand and wheel made and left plain, lacking sophistication. Three terracotta lids have also been unearthed. Wood charcoal and burnt clay have been noticed in all the levels. Significant discovery of this site is cultivated variety of carbonized rice in later phase of Neolithic period. Two post holes filled with charcoal, burnt clay, potsherds etc. have been identified by Sen as hearths. One probable hook and one weathered, brittle sickle were also among the discovery.

Polished edged stone axes are most commonly found Neolithic object over the site Dungi[81]. Other finds include chisels, hoes, scrapers, flakes, hammer stones, pounders, ring stones and large, heavy saddle querns etc. Few reddish brown and grey coloured potsherds have been found. They are hand made and mostly left plain except a few, which bear incised parallel designs.

Neolithic artifactual assemblage of Barudi include chipped stone adzes, axes, fragmentary animal bones, charcoal etc.[82] Hollow terracotta earthen ball, burnt earth, hand and wheel made pottery of varying colours and carbonised rice probably of later date have also been discovered here.

On the right bank of the river Sona, Jojo is located, where few celts were recovered from the gullies below the site.[83]

A sizeable amount of Neoliths have been encountered at Sini by D.Sen and A.K. Ghosh[84]. Neolithic finds include celts, chisels, ringstones, pounders and saddle querns. Potteries are both hand and wheel made. Carbonised rice has been found in a pit in association of a polished stone axe besides charcoal, burnt clay, bone fragment, iron object etc.

Ukri is another site on the Sanjai river which has yielded Neolithic celts, beads, microliths and pottery.[85] Crude hand made pottery have been found in association with stone celts, chisels, ringstones and other artifacts at Chakradharpur[86].

Bongara Neolithic assemblage consists of chisels, rounded butt axe, thick butt axes, wedges, perforated stones, hammer stones, faceted tools, and shouldered axe[87].

The Subarnarekha river served as a shelter of Neolithic folks in the state of Bihar also. On its bank a number of sites like Seraikella, Ghoranagi, Chandil, and Lotapahar have been located. From an exploration at Seraikella celts, pounders, saddle querns, hammerstones, chisels have been found[88]. Along the river valley at Ghoranagi Neolithic celts, mace heads have been found[89]. With other prehistoric remains, Chandil[90] has yielded a Neolithic celt. Two Neolithic celts and a polished roller have been found at Lotapahar[91] area near Chakradharpur. Other Neolithic sites of the Subarnarekha valley are Chakulia[92], Jamshedpur, Ghatsila[93] etc. Dora[94] near Dungi has yielded some Neoliths.

From the two sites of Dalbhum area, named Matigara and Ladup or Nadup, some peculiar hammerstones, besides one pounder and two arrow heads have been found[95].

Rohtas:

Neolithic tool kit of Akorhi[96] comprised of polished celt, stone hammer, rubber, pestle, fluted core, ring stone, terracotta animal figurine, discs, pottery etc.

Exploration at Malaon[97] revealed polished celt of basalt, ring stone and hammer. All discoveries of Sakas[98] have been made through surface collection. Stone industries of Sakas comprise of polished celts, flutted cores, blades and bladelets. Potteries of different shapes include footed bowl, vase, channel spouted bowl, lipped bowl etc.

At Daindih[99] Neolithic evidence like celt, rubbers, pestles, flutted cores, hammer stones and ring stones are found in association with black and red ware and N.B.P. wares. Recent explopation[100] in South Bihar have revealed a rich Neolithic culture particularly in Rohtas and Kaimur districts. Discovery includes small polished triangular celts, edge ground potsherds.

Ranchi:

Triangular wedge shaped axe, ring stone, flattish celt, arrow head and polisher were reported from Ranchi district[101]. Later on a narrow double edge celt was obtained by a zamindar from the village of Tappa Kakea[102]. V. Ball[103] found a beautifully made celt from a hill near Buradih village in Tamar. Here, Neoliths were collected with microliths.

In 1915 S.C. Roy reported[104] a large number of stone objects including stone celt from a number of sites which he called 'Asura sites'. About a hundred stone specimens including chisel, adze, triangular axe head etc. were found in Ranchi and Khunti area. Roy also described a highly polished hammer. A sporadic occurrence of Neolithic celts has been observed from Antigara of Ranchi district. A celt of schistose was picked up by Roy from Dargama. A. Ghosh collected few potsherds, beads and polished stone celt form Ite[105].

Santal Pargana:

Rev. P.O. Bodding[106] discovered about two thousand and one hundred forty nine stone implements like wedge shaped axe, perforated stone, mace head, shoulder headed celts, small oblong flat stone, adze, rubber, hammer from Dumka sub division of Santal Pargana (Plate 7; Fig.12). These collection was later on studied by F.R. Allchin[107] who sub divided the entire assemblage into following groups. 1) Indian axe, 2) Small celt, 3) Chisel and rectangular celt, 4) Shouldered celt, 5) Rubbing stone and 6) Hammer (Fig. 12).

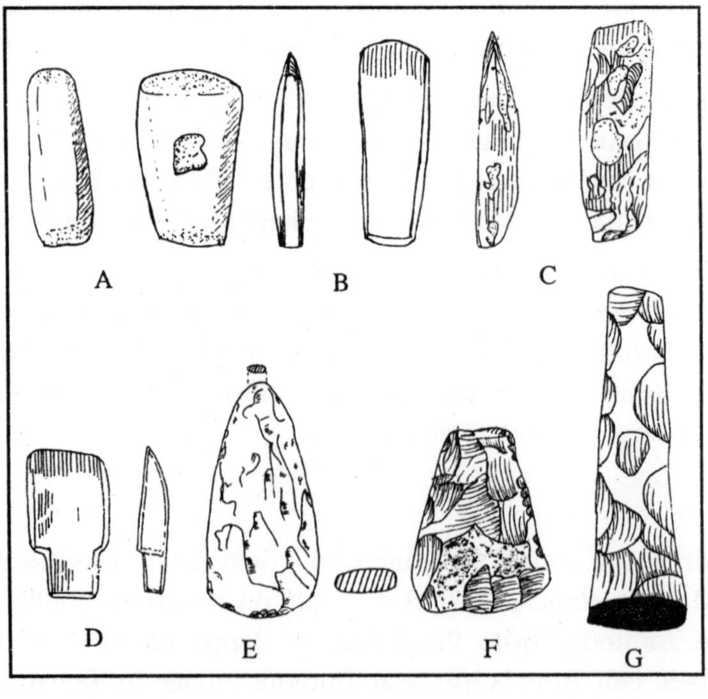

Fig.12. Neolithic Stone tools from Singbhum and Santal Pargana.

Gaya:

From Bodhgaya, a Neolithic implement was discovered by Knight Hallowes[108]. From the dry bed of the river Saraswati at the old site of Rajgir[109] recent exploration has brought to light celt with slight shoulder and a knife blade.

Stray occurrences of neoliths reported from other districts of Bihar are:

Only two Neolithic celts are reported from Jamalpur[110] region of the Monghyr district. One of them is unfinished and grooved for hafting and other shows a sharp cutting edge. In course of an exploration from Hatigara at Palamau[111] district, one Neolithic celt, hammerstone, pestle and gritty red ware, were found[105]. From another site at Akhra[112] a probable Neolithic artifact was picked up.

From Baragunda[113] at Giridih a polished ringstone has been recovered along with miscellaneous implements. A report on about twenty eight rockshelters containing tools of lower Palaeolithic to Neolithic and historical period from Nawadah and Girdih district was published[114].

So, it could be surmised that Neolithic finds of Bihar are more remarkable in both quality and quantity than those found from West Bengal. Stone age culture might have flourished in four different regions of the state i.e 1) Chhotanagpur plateau area II) South Bihar area III) Santal Pargana area and IV) North Bihar area. The north-western part of the state is archaeologically not very well haunted. In comparison to other parts of Bihar, the yield of Chhotanagpur plateau area are earlier, while in northern Bihar, it is more evolved with certain borrowings from the mid Ganges Neolithic traits. So, it could be said that from Chhotanagpur plateau stone age cultural tradition might have spread to other parts of the the state and north Bihar was the region to receive it last. The recently discovered neoliths from Nawadah and Giridih districts may throw useful light on Neolithic cave dwelling of eastern India.

ORISSA

The study of Indian prehistory has touched a century and it is true in case of Orissa also, but archaeological activities in this region were not very remarkable for a considerable period of time.In recent years, the study of prehistoric archaeology has gained a major impetus and a considerable number of Neolithic remains have been recovered (Map 3). Most of the prehistoric discovereies, here, are made through surface collection. In the year 1961, a stratigraphical position of eastern Neolithic culture was observed for the first time at Kuchai of Mayurbhanj district.

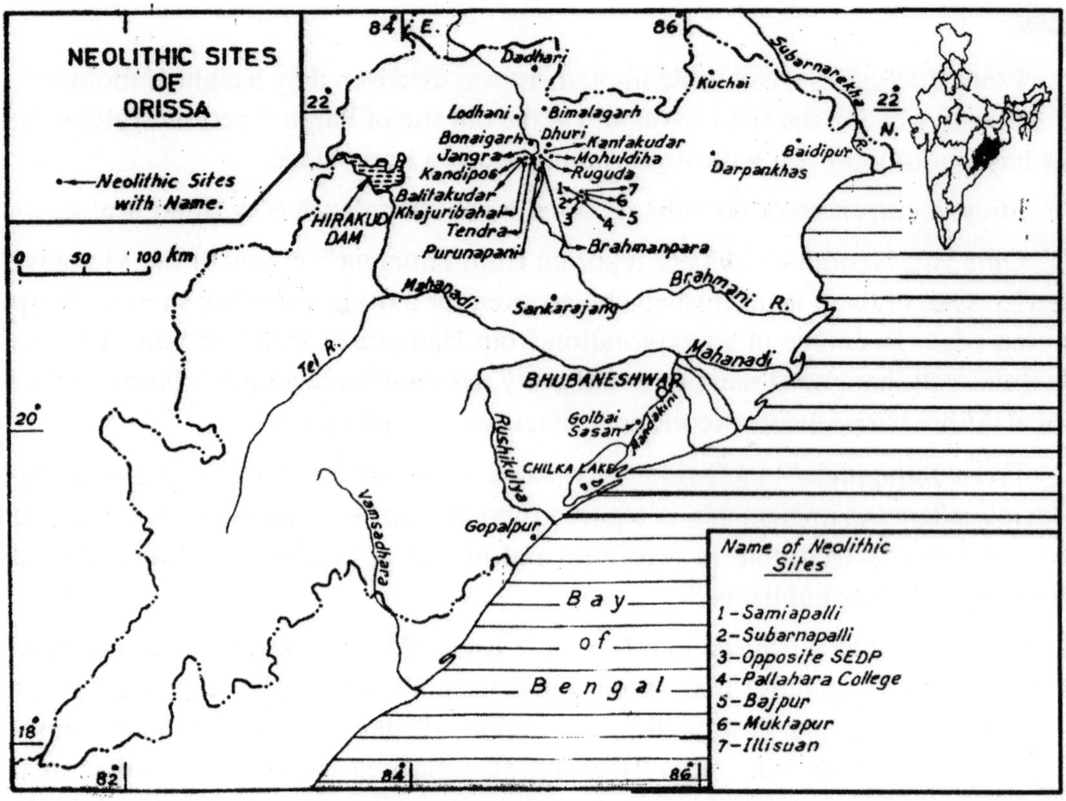

Map.3. Neolithic sites of Orissa.

Mayurbhanj:

Archaeological discoveries in Mayurbhanj are of very great interest. From this district a number of important Neolithic sites like Kuchai, Baidipur have come to light.

From Kuliana only 3 Neolithic ground and polished celts were found[115]. One of them has a pointed butt and square sides. First Orissan Neolith was collected from Sadar sub-division. Large number of Neoliths has been collected from Baidipur and neighbouring Pratappur and Madhupur area. The village Baidipur is situated on the eastern slope of the high ground in the south of the river Burha Balang. Neolithic assemblage of this site comprised of shouldered celts, adzes, conical hammerstone or corn crusher and pestle. Potteries were hand made or at best turned on hand lather and a few exhibit a fine red slip on them[116].

Jajpur:

In Jajpur district, six sites have yielded 27 Neoliths of various types recently.

Darpankhas, situated about 71km north of Bhubaneswar, has yielded neollithic tools like ground axe, chipped axe, adze, chisel, shouldered axe, arrow-head, and broken ringstone ((Plate 19). Other sites of this district are Sunamukhi tangar, Ranibandhi, Danmandal and Kahanikia tangar.

Sundargarh:

At Sulabdihi in Bonaigarh[118] sub division of this district thirteen Neolithic sites have been reported. Majority of the sites were celt manufacturing centres. In few sites celts were microchipped and were given final shape, while rest of the sites were habitational centres. Tool types were represented by a semi finished celt, axe, chisel, and pebble tools. Pebble tools occur in most of the sites and in a number of cases it outnumbered celt category tools (Fig. 3A-3C). About six neolithc implements were discovered from group of four caves known as Usha - Kuti in the river valley of Brahmani and Ib[119].

Deoposh[120] has been located on the river valley of Brahmani. One of the two trenches of this site has yielded countless waste flakes and few Neolithic celts. The other has yielded numerous potsherds of grey ware and coarse gritty red ware. At Dadhari about 8 km north of Birsa town[121], from some early historical mounds yielding pre N.B.P. ware and N.B.P. ware bowl, few Neolithic celts and ring stones have been found.

The upper Brahmani river bed has gifted the archaeologists with a number of sites like Riu, Misakund, Jakeikela, Garjan, Jangra etc.[122], whose yields are lower Palaeolithic, upper Palaeolithic as well as Neolithic in nature. Neolithic industries of these sites are represented by stone axe, chisel, ring stone, polished Neolithic celt etc. At Garjan and Jakeikela, a black burnished pottery was also encountered.

Berhampur:

Gopalpur is a Neolithic—Chalcolithic site in coastal Orissa. About 72 km south of Bhubaneswar, a small mound named Jagati, near Gopalpur has recently been identified for its prehistoric yield[123]. Exposed section of the mound contains celt, rubbing stone, grinding stone, perforated stone (Fig.3E) and burnt clay. Potsherds are mainly wheel made while some others indicate being made by hand as well as by turntable (Fig. 4J-4L). Ware comprised of red ware, red slipped ware, dull red ware, grey ware, cream slipped ware, chocolate ware, black and red ware and burnished black ware. A large number of animal bones have been recovered from this site ((Plate18). Out of a total of thirty seven bones, thirty one have been identified as of cattle, buffalo, nilgai, chital and wild pig.

Dhenkanal:

From Harichandanpur, both Palaeolithic and Neolithic tools have been picked up. Neolithic tools include a polished stone celt, a roughly finished chisel and a ring stone[124]. Explorers found two chisels, one ring stone and fragment of a polished stone celt from Kaniha[125].

Jimnia Hillock[126] is an open air site located in the upper valley of the Kumaria. Finds of this site include lower Palaeolithic and Neolithic ring stone and small celt. Neolithic finds of Sardarpur[127] are four chisels only. Sankeranjang[128] is also described as a Neolithic find spot.

About two hundred and eleven neoliths have been found from nineteen sites at Dhenkanal district [129]. The sites are - Nimidha, Sarang, Gotamara, Balaramprasad, Joragadia, Kurudole, Panchamahala, Turang, Ranigonda, Sakosingha, Phulapada, Talmul, Jamunali, Parang, Kuio, Sankerajang, Ghantapada, Balhar and Dera. Finds include axe, adze, chisel, bar-celt, scraper, ring stone, shouldered celt, chopper, earthen and stone bead, round and circular mud pellets.

Keonjhar:

From an exploration in Keonjhar district a number of prehistoric sites have come to light. These are Karadapal, Tikria, Noapara, Kachiladarah, Karadatani, Sanahudi, Palaspal, Chilikadarah, and Bhagmunda[130]. Finds are mostly microliths with few celts.

From Danguaposi[131] and its neighbouring area at the foot of a hillock was found three Neolithic celts with Microlithic core and waste flake. Mohanty[132] in his survey of Keonjhar district brought to light some fifty seven Neolithic sites like Champua, Ghasipur, and Ghatagaon etc. Neolithic finds include celt.

From Thakurani, celt of fine grained chert was reported[133]. Two chisels from Sitabhanji, old Keonjhar district have been found. A number of sites yielding about two hundred and five neoliths have come to light near Pallahara region[134]. On the right flank of the river Mankara at Samiapalli one axe, one chisel and a flake have been found. Subarnapalli, about 1 km away from Samiapalli has yielded a chisel and a flake. In Pallahara college area, on the embankment of a tank three Neolithic axes have been recovered among one hundred twenty prehistoric artifacts. Bajpur, about 1km from Pallahara college area finished and unfinished axe and chisel have been found. Few slow-wheel made coarse red ware and black and red ware potsherds are also found in course of exploration. Another site near Pallahara College area, opposite SEDP one Neolithic axe has been found from Muktapur and one unfinished axe, one flake and a rubber have also been collected. One finished and one unfinished Neolithic axe have been recovered from Illusion.

Stray occurrence

A number of celts of oblong section have been recovered from Jaugada in Ganjam district[135]. A single celt has been found at Malpadaghat near Shrimal village of Phulbani district[136] From a surface exploration, few stone celts have been encountered from Didwana, district Nagaur.

So, Orissa-which was not a favourite haunt of prehistorians, for a long, has witnessed greater degree of institutional involvement in archaeology, later. The Mahanadi river divided the state into two distinct Neolithic zones. North-eastern Orissa, which is a part of the Chhotanagpur plateau fringe area, is rich in bigger and flatter axes, smaller chisels and bar-chisels while the yield of southern part of the river Mahanadi has yielded smaller tools and ringstone variety. Kuchai is an important site of the former division, where a microlithic phase precede Neolithic phase. In southern Orissa, the stratified Neolithic and Chalcolithic level of Golbai sasan is one of the most intriguing discovery of eastern Indian archaeology. Evidence of cave habitation during Neolithic period is conspicuous in Orissa also.

ASSAM

Densely wooded hills and mountains caused limited field work in this state. Efforts were made by T. Bloch[138] and T.N. Ramchandran[139] to open up the area for archaeological activities. First Neolithic implement from this state was discovered from Darang district, near Tejpur. After that, Neoliths were found throughout the length and breadth of north-eastern India (Map 4). But the problem lies in the fact that majority of the finds come from the homes of tribal people and thus records of proof of such finds could not be achieved.

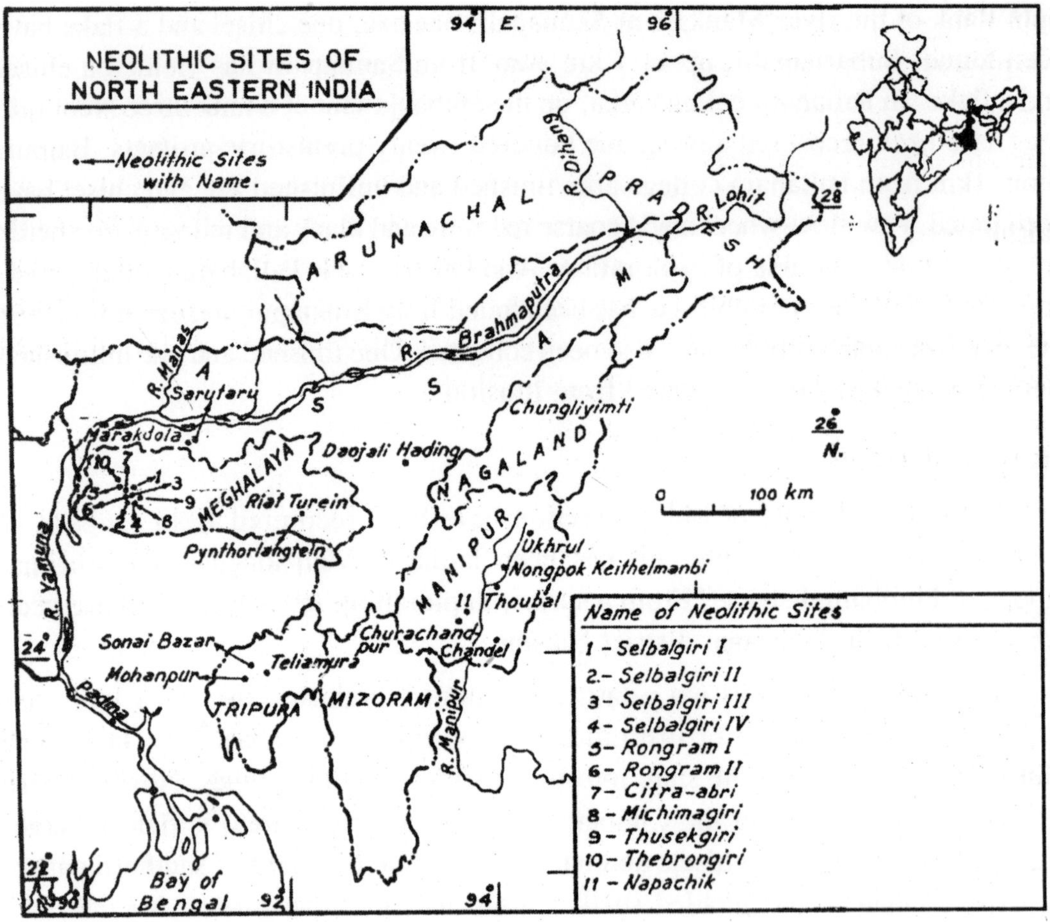

Map.4. Neolithic sites of North-eastern Indian states.

The present state of Assam can be divided into 3 physiographic divisions 1) The alluvial plains of the Brahmaputra valley in the north comprising the districts of Goalpara, Darrang, Lakhimpur, Kamrup, Nowgong, Sibsagar and Dibrugarh 2) The central Assam hills comprised of Mikir hills and north Cachhar hill district and 3) The hills and alluvial terrain in the south covering Cachhar district.

Goalpara:

Small Neolithic collection of this district comprises tanged celts, flat celts, adzes, rectangular and broad axes[140].

Darrang:

From an outlier of granite - geneissic rock at Biswanath[141] near Tejpur, Neoliths have been found which consist of celt, grooved hammer stone, grinding hammer stone and small and flat pebble.

Lakhimpur:

About Twenty two implements were collected from sites named Lahowal, Naharkatiya and Barkhamatigaon[142]. Tool types include tanged celt, flat celt, hog back type celt, chisel, quadrangular axe, miniature celt of jadeite, chipped celt, curved knife etc. From Rani[143] large axe, Neolithic chisel, small axe, miniature adze, shouldered celts were found. Kamakhya hill[144] has also yielded small axe, chisel etc.

Neolithic tanged celts, flat celts, small celts, rounded axes, chisels and adzes have been picked up from Loharghat, Tetelia, Sarania hills, Lunmati, and Sonpur[145]. Markadole,[146] about one km, from Sarutaru, yielded a shouldered celt, a terracotta object and wheel turned kaoline pottery of historical origin.

Sibsagar:

It is one of the largest districts of Assam but has yielded one tool, a small celt only [147].

Dibrugarh:

H.B. Medlicott discovered small celt of soft rock from two feet below the ground in plantation near Dibrugarh[148].

Following are few Neolithic find spots which are broadly located within a district or state but particulars of the sites are not given in the reports. J.P. Mills and J.H. Hutton described shouldered adze head and roughly shouldered triangular axe from north Cachhar hill district[149]. J.C. Brown described[150] about grooved hammer stones from Assam. H.C. Dasgupta also described[151] about two shouldered adze of Burmese type from Assam.

From trial digging at Langting - Haflong road at layer-1 celt, grinding stone, clay ball, cord impressed pottery, at layer-2 broken celts, grinding stone, similar potsherd of layer-1 and at layer - 3 shouldered celt, and cord impressed pottery have been found[152].

MEGHALAYA

Garo Hill:

In respect of Stone Age cultures, this zone is better known than any other parts of north eastern India.

First Neolithic tool from Meghalaya was discovered by G.D. Walkar from Garo hill district[153]. Dani described[154] some Neolithic sites like Fakhre Aap, Molmegiri, Adingiri, Rongeng, Dilmagiri, Rongkhongiri which have yielded facetted tool, round butt axe, axe with broad cutting edge and splayed axe.

Sharma[155] discovered further Neolithic findspots from this area. They are Rougchigir, Rambhagiri, Rongchongate, Holongiri, Rongjeng, Makra adap, Bongiri. Lithic industries of these sites include different types of celts, chisels, axes and hammer stones etc. In 1966-67 more Neolithic tools were collected from Matchakolgiri, Goudlgiri. Rongdu etc[156]. Rongdu has yielded more than three hundred tools and large number of potsherds.

Arbella range around Tura in Garo hills is the only area in whole of the northeastern region which has been most extensively explored. Repeated exploration has yielded most valuable evidence of prehistoric culture here. Important sites of Arbella range are Selbalgiri, Rongigiri, Thusekgiri etc.

The stone tool assemblage of the site consisted of 1) stone axe industry 2) flake and blade industry 3) micro blade industry and 4) ground stone industry. Selbalgiri locality has yielded stone axe and potsherd. Lithic industry of Selbalgiri locality 2 consists of handaxe, picks, discs, chipped axes, ground celt, shouldered celt, awls, scrapers, borers, blades, large knives, knife blades, points, hinge flakes, chisel, fabricator, spear head, arrow head, hammer stone, microlith with potteries from all the levels. Handmade potteries were devoid of any decoration. They were coarse, gritty in fabric and grey, grey brown or dull brown in colour.

Trial excavation at Rongram[157] 16 km. north of Tura, revealed Neolithic cultural level on the top layer. Important recovery of the site is broken piece of ring stone, a large number of celts of various shapes and potsherds.

About two hundred ninety-four Neolithic implements including celt, axe, shouldered hoe, adze, hammer stones, chisel have been collected from this area[158].

Citra-abri, a Neolithic site, about 1 km south of Rongru-abru, is located on the steep ridge on the left bank of the Rongram river. The Stone Age industry of the site could be divided into two broad groups *viz.* 1) flake blade industry and 2) Neolithic industry. Later comprised of axe, adze, hoe, chipped celt, scraper, point, borer, awl, spear head, shouldered celt, grinding stone, and a large number of potsherds. Potteries are handmade and left severely plain[159].

From Michimagiri[160] early, middle and late Stone Age tools have been discovered with Neoliths. Neolithic stone tool industry yielded chipped and ground celt, flake, blades, and arrow heads in association with crude hand made pottery. Most important finds of Thusekgiri[161] are chipped stone axe, flattish axe, crudely shouldered axe, cylindrical chisel etc.

Besides the above, Neolithic sites are also discovered from Thebrongiri which has yielded chipped celt, hammer stone, arrow point, knife blade, microlith etc[162]. From Dabakhol cave a solitary celt of fossiliferous limestone and from Gawak abri[163] both finished and unfinished artifacts have been found. Barapani has provided tools like adze, axe, point, scraper and waste flake besides finished and unfinished celts, microlith etc. From Mushak abri[164] ground and polished celt, crude hand made pottery and chipped celts have been discovered.

Khasi Hill:

This district could not prove itself as much potent in yield of neoliths. However, so far only six tools have been found - they are a curvilinear tanged celt, two quadrangular celts and two triangular celts[165]. H. H. Godwin Austen found one celt of hard slate from Maoflang at Khasi hill district[166].

Jaintia Hill:

In recent years two Neolithic sites have been recovered through trial digging[167].

At Pynthor langtein, trial trench has yielded a cultural deposit of Neolithic culturul milieu. Tool types are represented by adze, axe, chisel, point, blade, scraper, polisher, pen knife etc. (Plate 24). Ceramic industry comprised of hand made, cord impressed, coarse red ware sherds.

Trial digging at Riat Turein has yielded cultural remains of lower Palaeolithic to Iron Age. Layer - 1 is lower Palaeolithic and Layer - 2 is upper Palaeolithic and Mesolithic in nature. Layer - 3 is represented by Neolithic remains which consists of a single piece of Neolithic celt and potsherds of dull red ware.

ARUNACHAL PRADESH

Tirap:

As early as 1867, John Lubbock reported[168] the discovery of Neolithic celt from Namsang in Tirap district. Later on in 1870 three more celts were discovered from the same area[169]. Besides that, facetted tool, rounded butt axe, two long implements were discovered in course of time.

Lohit:

Stone Age sites were located at Chamba, Glow, Kale and Teehun of Lohit district. A few Neolithic tools were recovered from the area explored along the river Kamlang, Lati, and Tally[170]. Some shouldered celts, chisels triangular axes, and splayed axe were found from Dapabhum area. About seventeen Neolithic tools were collected during 1933-35 by J.P. Mills and J. H. Grace from the former Sadiya Frontier tract[171].

A number of sites adjacent to the Parsi - Parlo have been explored whose Neolithic yields are not less significant[172]. Vigo is located 16km south-east of Parsi-Parlo. One 'T'shaped (Fig. 7D) cylindrical hammer stone is found from this site. From Lia, 12 km south of Parsi-Parlo shouldered celt has been found. At Taram, 12 km north east of Vigo stone axe has been encountered. Double edged waisted axes (Fig. 7E) have been collected from Mindung site, 3 km west of Parsi-Parlo. 40 km south-west of Parsi-Parlo, at Taba rectangular axe, waisted blade, round scraper, hammer, chisel etc, have been found.

Siang:

One polished chisel and one stone adze are only Neolithic finds of this district[173]

West Kameng:

In early seventies, 18 Neoliths were collected from West Kameng district; they include edge ground axe and curvilinear shouldered celt[174]. One curvilinear rounded butt axe and a Neollithic celt were discovered from Mishmi hill area.

MANIPUR

Neolithic evidence is fairly widespread over both hills and plains. In 1969 a report on stray collection of Neoliths was published[176].

Ukhrul:

In 1969 form an exploratory trial digging at Khang Khai, Ukhrul district, a number of bone and stone artifacts and faunal remains have been collected[180]. About forty three neoliths from a surface collection were reported form Thongkhul, Ukhrul, and Thoubal area[177]. In 1979, from Tharon cave site edge ground pebble tools were discovered[178].

Churachand Pur:

From Churachandpur two Neolithic celts were reported[180]. Only one fully ground celt was reported by Sankalia[181] from a rich but unnamed Stone Age rock shelter of this district.

Imphal:

Phunan is another Neolithic site located 22 km south-east of Imphal. Neolithic celt having square cut sides, quadrangular axes, adzes, and hand made pottery with geometric pattern have been encountered[182] here. Association of these stone tools with pottery is not yet established.

Senapati:

The exposed section at Nongpok Keithelmanbi in Senapati district yielded ground stone tools of triangular hoe or axe type and cord marked potsherds[183].

Radiometric date of a charcoal sample from this site is 4,460 $\pm$ 120 years B.P. From Laimanai, near Kakching, tripod ware potsherds and Neolithic tools have been found[184].

Chandel:

During an exploration by state Archaeology department in Tengnoupal five Neolithic celts have been collected[185].

NAGALAND

Nagaland archaeology was mainly developed during the British regime. So, unlike modern reports the acutual locations of the Neolithic findspots were not reported at that time.

In 1867, a highly polished stone celt of greyish hue was collected probably from village of Namsang Nagas by H. Haly. E. H. Steel[186] also described another one of similar shape but smaller in size.

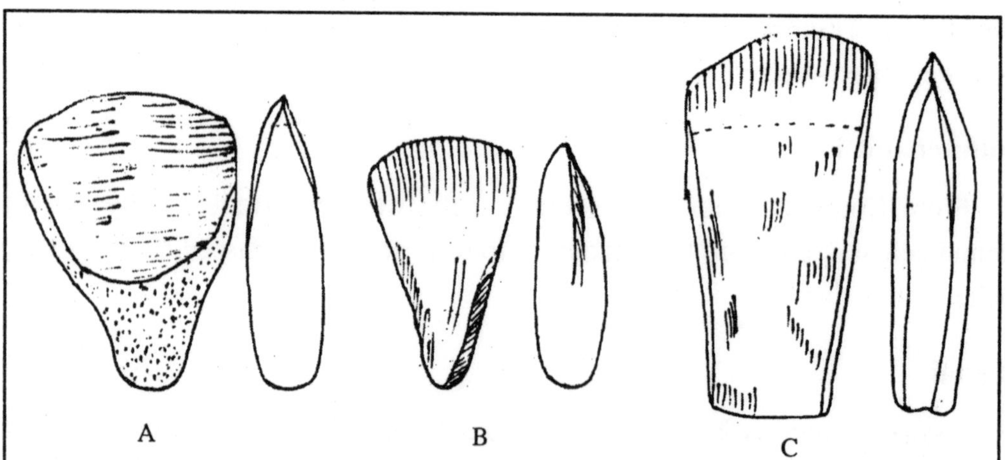

F. 13. Neolithic stone tools from Nagaland

In 1872 S.E. Peal[187] described a celt found 2 feet below the surface from a undisturbed stratified loam at Kanoo tea factory. Hutton[188] reported on different celts found from Naga hill region. One was roughly traingular with polished cutting edge and rudimentary shoulder. Another celt shows no shoulder while a smooth celt was described as having carefully cut square shoulders.

At Kigwema, 16 km south of Kohima, Neolithic celt, pestles and pounder were found at Chungliyimti in Tuensang district[189].

Mokokchung district have brought to light polished Neolithic celts, pottery, grinding stone, stone rubber pestle, stone ear ring from eroded surface[190]. Exploration in villages Phek Basa and Shatuza and Wazheho cement factory have yilded ground polished stone tool, some grinding stone and large number of crude, hand made pottery.

MIZORAM

Archaeologically, Mizoram is still very little known.One stone axe has so far been found[191]. It is a large and thin axe made on slate with two completely and one incompletely bored holes on the poles.

TRIPURA

During a geological investigation at Tripura in 1980-83, about half a dozen rich implementiferous sites in the Khowai and Haora valley were located in West Tripura district.

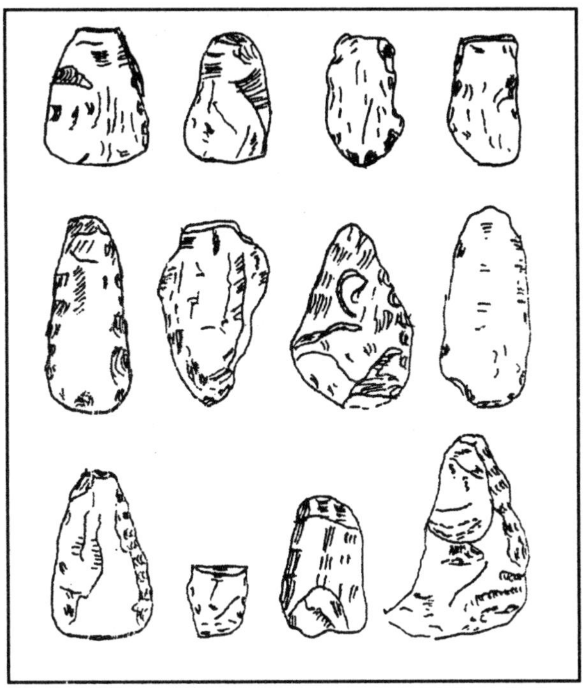

Fig. 14. Neolithic tools from Tripura.

West Tripura:

About seven hundred stone implements have been collected from sites like Teliamura, Jirania, Sonai Bazar, Mohanpur, Bishalgarh, Agartala etc.[192] (Fig. 14).

TeliaMura cultural materials include ground stone axes, adzes, points, scrapers, points, grinding stones and pebble halves. From five rich sites of Sonai Bazar area Neolithic adzes, ground axes, tiny impressed potsherds, hammerstone and fossil wood artifacts were found along with other pre Neolithic cultural objects. Neolithic artifacts collected from Sonaram area include ground stone axes, adzes, potsherds etc.

From distribution pattern of Neolithic sites over the present study area, it could be deduced that in river valley area, general difficulty was to obtain stone for manufacturing tool. Thus hilly area sites like Taradih, Sonepur etc. have yielded stone tools but habitants of river valley sites like Chirand, Chechar relied more on bones of their hunted animals for making tools. Now the survived objects of stone, bone, wood, clay etc. will be described along with remains of flora and fauna from our study area.

Stone tool :

Shouldered tool: This variety is most common in north-eastern Neolithic zones, though Bengal, Bihar, Orissa have also yielded this tool.

Splayed axe: This group of tools is found from Cachhar hills and Garo hill zones mainly.

Rounded butt axe: This tool type is found all over eastern and north eastern Indian Neolithic sites.

Axe with broad cutting edge: It comes mainly from Garo hills area, besides Ranchi, Santal Pargana districts of Bihar

Tanged axe and wedge blade : This type is known from Naga hill zone, Assam, Bihar and Bengal.

Wedge shaped axe: This type is found from Bihar.

Flat Celt: This type has been reported from the Garo hills besides Kamrup, Goalpara, Lakhimpur districts of Assam and Ranchi district of Bihar.

Double edged celt: This type of celt has been described by Wood Mason from Chhotanagpur plateau.

Bar-Celt: From Bon-Asuria of the then west-Dinajpur this type has been reported.

Quadrangular adze: This tool has been found from north Cachhar hill and Kamrup district. In Manipur also few specimens have been reported.

Ring stone: It is found from almost all over the study area.

Microlith :

Microlith which is commonly occurred in association with objects of different cultural period, is also found from a number of sites of the present study area. Mentionable sites, yielding microlith in association with Neolithic materials are-

WEST BENGAL

Tarafeni river valley (District Midnapur)

About fifteen different types of microliths have been found from this river valley area. They are fluted core, parallel sided blade, obliquely blunted blade, point, burin, lunate, end scraper, side scraper etc (Fig 11).

Bhairavbanki river valley (District Bankura)

About twenty microlithic artifact type including fluted core, various types of blade, lunate, discoidal core, burin, point, knife, triangle etc (Fig.11) have been found from this site.

Besides these, few microliths have also been found from a number of sites in West Bengal, Kankrajhor, Kattara, Ganganirmath in Midnapur district, Kurkutia, Susunia hill in Bankura district and Besra in Purulia district.

BIHAR

Chirand (District Saran)

Microlithic tool type of this site includes parallel sided blade, scraper, serrated point, lunate, borer and few geometric microliths.

Taradih (District Gaya)

From this site micro blade, point, disc etc. have been found.

Maner (District Patna)

From Maner microlithic tool like parallel sided blade, lunate, bladelet, etc. have been found in Neolithic level (Plate 16B)

Beside these, from Ukri in Singhbhum district microlith is found along with Neolithic artifacts.

In Orissa and north-eastern Indian states occurrence of microlith in association with Neolithic materials is comparatively rare. Excavated sites of Kuchai (Orissa), and Selbalgiri (Meghalaya), microliths are found in pre-Neolithic context. From some lesser known explored sites like Riu, Misakund in Sundargarh district, Orissa and Barapani in Garo hill district, Meghalaya have yielded some microliths along with Neolithic artifacts.

Bead:

Beads of shell, semi precious stones have been found from both from southern and northern Neolithic zones of India. In eastern India very few sites have yielded evidences of beads in its pure Neolithic level.

Chirand (District Saran)

A good number of beads with beautiful finish have been unearthed from excavation. They are of tubular, barrel, cylindrical, triangle and disc shaped and made of rocks like chalcedony, marble, jasper, agate, steatite, faience etc.

Bongara-Bhangat (District-Singhbhum)

Three very small beads one each of spherical, cylindrical and disc shape have been found from this site. The raw materials of these beads were not reported.

Besides these sites few beads have been found mentioned from Ukri (Singbhum district), Senuwar (Rohtas district), Maner (Patna district) and Buradih, Ite (Ranchi district).

Other states of eastern and north-eastern India could not provide any evidence of Neolithic beads.

Bone and Antler Tool:

The term prehistoric cultures are generally used as synonymous of stone using cultures but as bone is easy to work upon and hard it could be assumed that its products might be an important constituent of the man's tool kit. The use of bone tool could be traced back from the beginning of human activity, but as bone is easy to disintegrate most of it might have been lost to us. In some Indian Neolithic cultures *viz*. Chirand, Burzahom bone tools played an important role in their material cultural life.

WEST BENGAL

Bhaluk Sonda cave (District Bankura)

Although it is yet to determine the definite chrono-cultural horizon for the industry of Bhaluksonda cave, the whole Susunia hill (West Bengal) area has yielded large number of sites belonging to early stone age, middle stone age, late stone age and new stone age category tools.So, its bone tool industry has also been considered here. The bone tool assemblage of the Bhaluksonda cave are needle, worked bone, scraper, double pointed perforator, spatula, single barbed harpoon head, socket, socket with hole, awl, bodkin, arrow head, worked bone etc.

BIHAR

Chirand (District Saran)

In eastern and north-eastern India, bone tools are known mainly by an impressive collection from Chirand. Bone tools from Chirand excel both in quality and types from that of Burzahom. Total one hundred and fifty bone tools include chisel, burnisher, bar-celt, hammer, wedge, scraper, shaft straightener, punching tool, javelin head, tong, borer, awl, needle, file, compass, arrow head, cutting tool, drill, spear-point, disc, pointed pencil like tool, surveying instrument, socketted handle, socket, adze, antler piece with cut marks and cattle shouldered blade. Along with this, ornament like bangle, pendant, ear-ring and comb are also found (Plate 12).

Senuwar (District Rohtas)

Bone tools of Senuwar could be divided into three categories *viz.* finished, partially finished and unfinished. Among them though unfinished specimen should not come under the category of tools, till micro wear study reveals the various operations like scrapping, piercing, punching etc. by the naturally broken part of bones. The partially finished and unfinished bone tools are frequently found in Neo-Chalcolithic level of Senuwar.

Golbai Sasan (District Puri)

In Golbai Sasan, at Orissa bone tools were recovered in good number from its Chalcolithic level but in Neolithic level no bone tools were met with. In Neolithic stratum two pieces of bones with working marks on them were found along with some partially mineralised bones and anlters.

With stone and bone, most easily available wood must have been used by prehistoric man. As anything made of wood, bamboo, grass or such perishable material vanish into dust within a few years or centuries so it is not possible to get wooden implements from prehistoric context. But it is remarkable that the tool industry of Tripura is almost wholly based on silicified fossil wood. In the absence of true hard rock in Tripura, it could be assumed that, silicified fossil wood was extensively used for making implements. Broad tool types of Tripura Neolithic industry are various types of celt, scraper, point, bifacial chopping tool, hammer stone, fluted blade cores, limace (slug), pick axe and grinding stone of which 98 percent are made up of fossil wood.

Pottery:

As it is emphasised for a number of time that stone or bone tools are not the only resource of the prehistoric past. Along with these survivals, pots and pans of clay could also stand the tear and wear of nature and constitute an important element of prehistoric archaeology. In this study area evidence of Neolithic potteries comes mostly from sites of Bihar. Besides that few sites from states of Orissa, Assam, Maghalaya and Manipur have been reported.

WEST BENGAL

In West Bengal very few potteries have been found.

Kurkutia (District Bankura)

Pottery from Kurkutia includes a few hand made, coarse, grit tempered sherds inspite of heavy weathering, a few of them show some slip on both the sides. Besides that some unique examples of sand stone (stone ware) were collected from this site.

BIHAR

Sini (District Singhubhum)

The Neolithic pottery of Sini was both wheel turned and hand made. They are generally grey, black and brown to red in colour.

Borda (District Singhbhum)

Local clay in chopped straw or chaff as tampering material was used in manufacturing the pottery of Borda. The dull, lustreless brownish surface of the pottery show slightly reddish tint. This group of potsherds is more or less porous and hand made.

Bongara (District Singhubhum)

Semi-lateritic, local residual clay is used as the raw material of this group of pottery. Texture of this clay is very thick and coarse and full of lateritic and carbonaceous particles. The surface is dull light brown, which proves that they were mostly hand made.

Dumdumi (District Singbhum)

In the composition of clay of Dumdumi we find a proper mixture of fine residual clay with sand. The outer surface is deep brown in colour with reddish tint. The surface texture of all the sherds is smooth and compact with a slight degree of variation. The slipped surface shows a certain degree of lustre. Uniform, parallel circular marks reveal their manufacturing process.

Chirand (District Saran)

Pottery from phase 1A of Chirand is mostly hand made grit- tempered red ware. The clay is mixed with mica and quartz particles. Potteries were represented by vase, bowls, pot with spout etc. (Plate 14A, B, Fig.1). Most of the pots were burnished either on exterior or interior or on both the sides. Thin grey slip was often applied before burnishing grey pots. They were painted with ochre colour after firing. Applique decoration was found on dull red ware.

Senuwar (District Rohtas)

In Senuwar from period I pure Neolithic level, three main ceramic types *i.e.* Red ware, burnished red ware, burnished grey ware are encountered. Besides that rusticated ware and cord impressed ware are also noted. Another mentionable yield of ceramic industry of Senuwar is mainly wheel made though hand made potteries were also quite extensive but less than wheel made variety (Plate 9).

Chechar – Kutubpur (District Vaishali)

The pottery of period-I of Chechar Kutubpur is, in general, made of well levigated and fine to medium grained clay. It is made on wheel and well fired. A thick and fine slip ranging from chocolate through crimson red, light red, orange to ochre colour is usually applied. Grey ware is represented in tones of deep grey to blackish grey. Pottery surfaces were mostly burnished, a distinct feature of the ceramics of this period.In period IA and IB red and grey ware are dominant while in period 1C grey, black and red ware increased.

Taradih (District Gaya)

Pe IA of Taradih ceramic industry is represented by hand made, slipped and burnished pottery. Period IB pottery is hand made, burnished grey ware. Few sherds are produced by turn-table method. Cord impressed sherds are also evident in this phase.

Maner (District Patna)

Neolithic potteries of Maner are both hand made and wheel made. Besides these two techniques turn table method was also known to the folks. In the ceramic assemblage of Maner, red ware and burnished ware predominates over the burnished grey ware and pale grey ware.

ORISSA

Kuchai (District Mayurbhanj)

Two types of fabric have been found in Kuchai Neolithic pottery. This pottery has been termed as coarse grit tempered red ware. The thicker pots were invariably hand made while the thinner variety was made by turn table method or slow wheel method. Exterior surface of most of the pots are red slipped (Fig.4). Similar types of pottery have been found from Baidipur and Khiching of the same district.

Golbai Sasan (District Puri)

Period-I of Globai is dated to Neolithic phase. Pottery of this period is hand made with cord and reed impression. Some sherds however seems to be wheel made (Fig.4).

Gopalpur (District Khurda)

Exposed section of the Jagati mound near Gopalpur contains a number of potsherds. These are mainly wheel made and some others indicate being made by hand as well as by turn table. From this Neolithic Chalcolithic site red ware, red slipped ware, dull red ware, grey ware, cream slipped ware, chocolate ware, black and red ware, and black ware have been collected (Fig.4).

ASSAM

Daojali hading (District North Cachhar hill)

Ceramic find of Daojali hading have been divided into four categories on technological ground. They are (i) cord-marked pottery (Plate 22) (ii) incised pottery (iii) stamped pottery (iv) plain fine red ware. Among them cord marked and incised designed potteries are generally grey in clolour while plain fine red ware is brick red in colour. Other shades are dull red and chocolate brown. Among the four types, the first three are made of coarse and impure clay while for the last one, fine clay was used. Vessels were hand made except for that of plain red types, which might be turned on a pivoted turn table. In some cases coil or ring building method was used for heightening the wall of the pot. Parallel grooves in rows, double threaded twisted cord marks were common decoration. Imperfect firing results in dark core which is evident in almost all cases.

Sarutaru (District Kamrup)

Sarutaru Neolithic pottery does not reveal any complete shape. The whole set of pottery is hand made, coarse, gritty and unslipped. On the whole the ware is unrefined and not well fired. Ceramic types include brown, buff and grey wares. The ceramic is sometimes decorated with cord impression or basket impressisons (Fig.6).

MEGHALAYA

Selbalgiri (District Garo hill)

The pottery found from Selbalgiri is hand made, coarse and gritty in fabric. Shade ranges from grey brown to dull brown with a very rough texture due to the presence of quartz grits in the clay. The potteries are plain and without decoration.

Rongru Abri (District Garo hill)

A large number of potsherds have been found from a partial search of cliff and slopes of Rongru Abri. By and large, the pottery of this area is hand made. Grey coloured pottery dominates the collection. An impure clay tampered with quartz grits seems to have been employed for the manufacture of pottery.

Riat-turein (District Jaintia hill)

From a trial digging few potsherds of dull red ware have been found from layer-III.

ARUNACHAL PRADESH

In Arunachal the only excavated site Parsi parlo has yielded evidence of potsherds.

Parsi-parlo (District Subansiri)

Small numbers of sherds have been found from phase-II of this site. These hand made sherds represent square grid and honey comb beaten impressed designs.

MANIPUR

From the Neolithic culture of Manipur three different pottery types have been found (Plate 26)

Nongpok Keithelmanbi (District Senapati)

The pottery type found from this site is called corded ware. This variety is hand made, ill fired and decorated with cord marks. In the absence of any other Neolithic character with this corded ware pottery it has been suggested as a sub-Neolithic type as in other parts of South-East Asia.

Napachik (District Bishnupur)

The pottery of this region is characterised by the cord marked pottery with tripod legs. The tripod ware pottery might be used for cooking purposes also.

Phunan (District Imphal)

Pottery from this site is handmade and decorated with geometric pattern. The patterns of this geometric ware consist of circle, parallel lines, wavy or straight lines either in pairs or band etc. which are produced by incise, applique and stamped methods.

Terracotta :

Terracotta objects are another mentionable discovery of prehistoric culture. In eastern and north-eastern Indian Neolithic culture terracotta objects have been found from Bihar only with one site from Nagaland.

Followings are few noteworthy discoveries-

BIHAR

Sonua, Barudih (District Singhbhum)

Terracotta lid has been found from Sonua. Terracotta ball has been reported from Barudih.

Senuwar, Akorhi (District Kudra)

Terracotta figures have been discovered from Senuwar and Akorhi.

Taradih (District Gaya)

Terracotta balls have been found from Taradih.

Chirand (District Saran)

From Chirand terracotta bangles, toy cart wheel, perforated terracotta disc etc. have been found from its Neolithic level.

Maner (District Patna)

Terracotta bead, ball, spindle whorl etc. have been found from Maner.

NAGALAND

Tuensang district.

Terracotta object has been reported from Tuensang district of Nagaland.

In addition to the above material relics of Neolithic period few fossiliferous floral and faunal evidence contribute greatly to the archaeological reconstruction of this region.

In the post Pleistocene variegated environmental mosaic, agriculture might have been initiated. The theory of Western Asia as the birth place of cultivated plants remained anchored for sometimes. But during the last three decades field work in areas outside the *fertile crescent* have offered alternative to this postulate.

It has been already discussed that scholars have identified South-East Asia including India as the most potential zone in respect of origin of agriculture. She has a net of rivers and ranges of mountains, wet monsoon climate suitable for the development of agriculture. It is generally believed that nature has destined her to have thriving agriculture. The non availability of direct archaeological evidence in considerable number and differential preservation of pollen make it a highly puzzling and complicated issue.

In assessing the position of agriculture during the Neolithic stage in eastern and north-eastern India, it could be quoted from Randhawa that 'when one examines rice cultivation in India at present, one finds that its largest concentration is in eastern India *viz.* West Bengal, northern Bihar and Orissa'[193].

In case of West Bengal, Chakrabarti has claimed that 'the roots of beginning of food production and animal domestication in this region should be looked for in the microlithic context than in any hypothetical Neolithic situation'[194]. P. Sinha traced signs of plant utilization along with the microlithic artifacts of Bankura[195] but the work is not published in detail.

According to Chakrabarti 'In Bihar, at Santal Pargana, such imponderables comes from the Rajmahal hill area. At Panchrukhi, a handmade heavily weathered buff red ware was found among microliths. Although its date is not certain but the fact is suggestive of an early adaptation to agriculture in that region[196].

In Orissa, a thermoluminescent date of 1,000 B.C.E. is suggested for Kuchai Neolithic pottery. A similar assemblage from neighbouring Baidipur has yielded pollen evidence of cultivated rice. Besides that growing of wild rice in Koraput - Jeypore tract was known long back. Recent studies reveal presence of domestication of rice in mesolithic level of Bajpur.

Plant Remains :

BIHAR

Chirand (District Saran)

At Chirand from lowest level (layer 17) above natural soil wheat (*Triticum sphaerococcum),* barley *(Hordeum sp),* wild rice *(Oryza rufipogan, Oryza perennis),* legumes *(Pisum arvens, Lathyrus sativas, Lens culinaris),* black gram *(Phaseolus mung)* and extremely small seeds (may be weeds rather cultivars) along with some unidentified fruit seeds have been found.

Senuwar (District Rohtas)

At Senuwar also from lowest level above the natural soil carbonized grains of different crops have been found. They are barley *(Hordeum vulgare),* field pea *(Pisum sativum),* lentil *(Lens culinaris)* and some millets. Besides that dwarf wheat (*Triticum sphaeoroeoccum*), grass pea *(Lathyrus sativa)*, Kodon *(Paspalum scrobiculatum),* vetch *(Vices sativa)* grew during winter season. In the following

Neolithic-Chalcolithic layer bread wheat *(Triticum aestivum),* chick pea or gram *(Cicer arictinum),* green mung *(Vigna radiata)* are encountered.

Extraneous influence of cultivation of Plants can be refuted by the presence of indigenous species of rice at Senuwar and Lahuradewa. On this basis it could be said that in the Vindhyan-Belan region, independent domestication event gave birth to an distinctive Neolithic culture, which is independent of other countries.

Animal Remains :

BIHAR

Chirand (District Saran)

From Chirand humped cattle *(Bos indicus),* Indian domestic buffalo *(Bubalus bubalis),* domestic sheep *(Ovis aries dolichura)*, domestic pig *(Sus scrofa cristatus)* domestic dog *(Canis f*amiliaris)[197] have been found.

WEST BENGAL

Bharatpur (District Bardhaman)

From Bharatpur domestic pig *(Sus scrofa),* domestic fowl *(Gallus sp*); pariah dog *(Canis familiaris),* buffalo *(Bubalus bubalis),* humped cattle *(Bos indicus),* domestic sheep *(Ovis orientalis),* domestic goat *(Capra hircus*) are discovered[198].

Besides the above mentioned domestic species of animals, wild species of spotted deer *(Axis axis),* swamp deer *(Cervus duvaceli),* Indian elephant *(Elephas niaximus),* one-horned *rhino (Rhinocerous unicornis),* from Chirand and horse *(equus cabellus),* swamp deer *(Cervus duvaceli),* spotted deer *(Axis axis),* barking deer *(Muntiacus munyak),* Nilgai *(Boselaphus tragocamelus)* have been found from Chalcolithic level of Bharatpur, West Bengal. From Dhuliapahar of West Bengal extinct cattle *(Bos namadicus),* black buck *(Antilope cervicapra)* and spotted deer *(Axis axis)* have been found. From Susunia, porcupine *(Hystrik bengalenis),* extinct cattle *(Bos namadicus),* elephant *(Palaeoloxodon namadicus)* have been found. From Bhaluksonda cave humped cattle *(Bos indicus),* wolf *(Canis linu)* bones have been identified and from Devapahar cave barking deer *(Munticits munyak),* sambar deer *(Carvus unicolor),* dog *(Canis familiaris),* cattle *(Bos indicus)* and river turtle

(Trionix gangeticus) have been found. From Golbai sasan animal remains unearthed from different level are humped cattle *(Bos indicus),* Indian elephant *(Elephas maximus),* domestic goat *(Capra hireus).* From Neolithic-Chalcolithic site of Gopalpur in Orissa out of thirty seven, thirty one animal bones have been identified as belonging to cattle *(Bos indicus),* buffalo *(Bubalus bubalis),* nilgai *(Boselaphus tragocamelus),* chital deer *(Axis axis)*,and wild pig *(Sus scrofa).*

So, it can be concluded that in the Neolithic culture mankinds's great leap towards advancement preserved in these archaeological evidences that suggest a comparably elaborate pattern of livelihood in that remote past.

Stratigraphy

Few stratigraphic records available in the study area, are listed here

WEST BENGAL

Midnapur:

Tarafeni River site

Silt and Clay

Loose gravel

Silt, clay and nodule

Gravel

Silt and clay

Cemented gravel

Bed rock

Dhuliapaliar

Recent

Semi compact brown earth - iron slag, iron pin, dull red and black ware potsherds

Fine grey and yellow earth -sterile

Semi compact yellowish sandy earth - Neolithic ground celt, hand made bowl of pale black, pale red, grey ware.

Compact yellowish grey earth - sterile.

Coarse ground sandy earth with nodules - Mesolilthic implement

Sterile layer

Basal rock

Pebble deposit - middle and upper Palaeolithic implements.

Bamandiha

Humus

Compact sticky hard grey earth with lime concretion and sandy patch.

Ferruginous concretion.

Cemented gravel - middle Palaeolithic digging point found.

Compact clay with gravel and boulder with sticky earth.

Thick laterite soil.

Susunia region, Bankura

Humus

Grey soil

Yellow loamy soil with calcareous nodule and laterite pellets.

Mesolithic cultural level.

Compact brown earth with lateritic pellets and ghooting.

Layer containing calchi indicate dry phase.

Boulder bed.

Bedrock or granite gneiss

Birbhanpur, Bardhaman

Light brown sandy earth.

Lateritic gravel with microlith, stone fragment, coarse granule

Mottled silty sand.

Bedrock of decayed sand stone.

Birbhum

Yellowish sandy deposit.

Compact lateritic matrix — Middle Paleolithic / Lower Paleolithic

Mottled clay - non implementiferous.

Primary laterite.

Bedrock.

BIHAR

Sonua nullah, Singhbhum

Reddish brown soil.

Unconsolidated gravel.

Silty clay.

Consolidated gravel.

Archaean schist

ORISSA

Mayurbhanj, Confluence of the river Baitarini and Khadkei

Surface and dry bed of river- microlith and polished stone celt.

Fine gravel -handaxe, cleaver, scraper, chopper, blade, burin etc.

Red silt -non implementiferrous.

Coarse gravel - handaxe, cleaver, scraper, chopping tool.

Mottled clay - non implementiferous.

ASSAM

North Cachar

Blackish brown clayey natured earth mixed with pebble.

Clay, sandy earth of brownish colour - broken celt

Reddish yellow soil of medium texture and sandy in nature - cord impressed pottery, shouldered celt,

Sterile layer.

Daojali hading

Layer - I Dark loose earth 8" - thick archaeologically sterile.

Layer-2 Dark grey and loose upper half and light grey lower half 2.5' thick, bearing stone tools (neolith) and potsherds.

Layer- 3 Yellowish brown compact earth.

In the study area at least four types of settlements could be observed, *i.e.* 1) mountainous 2) rugged hill 3) river valley and 4) coastal.

It could be naturally assumed that these mountainous or rugged valleys were not habited during winter season. Besides that the thickness of cultural layer of mid Pleistocene time also suggests that these high altitudinal sites might be temporary camp sites of favourable season. In the winter season they might have migrated to lower river valley or lake areas, where various aquatic foods were available in addition to fertile land products. Ultimately they preferred to settle permanently in this more facilitated surroundings. Archeaological and osteological evicences suggest seasonal migration of Vindhyan Mesolithic folks to other places of the Ganga basin. In this plain land some sites like Taradih, Sonepur, Chirand, Chechar Kutubpur, Golbai Sasan etc, have yielded Neolithic cultural deposit of more than three metres, suggesting towards a comparatively sedentary livelihood.

References:

1 Davidson, A. 1788, On some Roman Coin Found at Nellore, *Asiatic Researches,* I, pp. 145-170.

Goldingham, 1795, Some Account of the Cave in the Island of Elephanta, *Asiatic Researches,* IV, pp. 409-417.

2 Sankalia, H.D. 1974, *Preshistory and Protohistory of Indian and Pakistan*, Poona, P. 12.

3 Le Mesuriar, H.P, 1861, Note on Twelve Stone Hatchet or Celts from the Neighbourhood of Jabbalpur, *PASB,XXX,* pp. 81-85.

4 Wheeler, R.E.M. 1947, Brahmagiri and Chandravalli: Megalithic and Other Cultures in Chitaldurg district Mysore, AI, no. IV

5 Walsh, E.H.C. 1904, A Note on Stone Implement Found in the Darjeeling District, *MSB,* pp.20-24.

6. Verma, B.S. 2007 Chirand Excavation Report 1961-64 and 67-70, Patna.

7. *IAR,62-63,P.6*

8. *IAR,* 77-78, pp. 17-18

9. *IAR,* 84-85, P. 26

Prasad, A.K. 1980-81, 'Excavation at Taradih', *Puratattva,* no. XII, pp. 138-139.

10. *IAR,* 84-85, P. 12-13

11. Bose, N. K. 1928, 'Prehistoric Research in Mayurbhanj', *SC,* VI(ii), pp. 14-16

12. *IAR,* 87-88, P. 35

Sinha, B. K, 1998-99, 'Excavation at Golbaisasan. *Puratattva* XXIII, pp. 48-50.

13. Rao, S.N. 1973, 'Excavation at Sarutaru : A Neolithic Site at Assam' *ME,* 1, pp. 1-9.

14. *IAR,* 1 962-63, P.3

15. Sharma, T.C. 1970, 'The Stone Industries of Selbalgiri', *PISC,* (57th session) Pt. III, P. 485

– 1974, 'Archaeological Discoveries in North-Eastern India, *Puratattva.* VII pp. 17-19.

16. Ashraf, A.A. 1990, *Prehistoric Arunachal: A Study on Prehistory and Ethnoarchaeology of Kamla Valley,* Directorate of Research, Arunachal.

17. Singh, O.K. 1983, *Archaeology of Manipur, Napachik: A Stone Age Site in the Manipur Valley,* Govt. of Manipur.

18. Sankalia, H.D. 1974, *Op.cit,* p. 263

19. Walsh, E H. C. 1904, *Op.cit,* pp 20-24.

20 *IAR,* 62-63, P. 43

21. *IAR,* 66-67, P. 45

22. *IAR,* 75-76, P. 57

23. Goswami, K.G. 1948, *Excavation at Bangarh, Calcutta. P.32*

24 Chakraborty, M., 1967, Occurrence of Neolithic Implement in West Bengal, *Bull. CRI,* VI (1 and 2), P. 103

IAR, 64-65, P.46

25 *IAR,* 83-84, P. 48

26 *IAR,* 75-76, P. 58

27. Chakrabarty, M. 1967, *Op.cit*, P.99

28 *IAR, 64*-65, P. 48

Datta, A.K. 1988-89, 'Earliest Evidence of Painting in Bengal', *IMB,* XXIII, pp. 11-13.

29. Datta, A.K. 1992, *Neolithic Culture of West Bengal*, Delhi, P.59

31. Bhattacharya, M. 1983, A Note on Recent Exploration in the Upland Bengal in R.K. Sharma (ed) *Indian Archaeology - New Perspective,* pp. 124-132

32 *IAR,* 64-65, P. 46

33. Chakrabarty, M. 1967, *Op.cit,* P. 101

34. Chakrabarty, M. 1 967, *Op.cit,* P. 101

35. *IAR,* 89-90, P. 36

36. *IAR,* 83-84, P. 64

37. Chakrabarty, M. 1967, *Op.cit,* P. 101

38. *ibid* P 103

39. *ibid* P- 103.

40. Mukherjee, T. 1958, Neolithic Implement from Tamralipta, *Indian Folklore,* 1 (iii), pp. 42-44.

An Rep. ASI, 1921-22 P.74 ff.

41. *IAR,* 75-76, P. 58

42. *IAR,* 83-84, P. 94

43. Sen, D. 1948, A Note on Some Celt and Chisel from West Bengal, *SC*, XIV(vi), pp. 252-253.

44. *IAR,* 55-56, P. 69

45. *IAR,* 77-78, P. 86

46. Chakrabarti, Dilip K. 1990, *Archaeology of Eastern India,* Delhi, P. 172.

47. Ghosh, A. 1962, Prehistoric Stone Implement from Midnapur, *SC*, XXVIII (vii), P. 338

48. *The Telegraph*, 18th Apr. 1995, P. 1

49. Dasgupta, P.C. 1966, *Archaeological Discoveries in West Bengal,* Directorate of Archaeology P. 27.

50. *IAR,* 57-58, P. 69, 59-60, P. 48, 65-66, P. 58.

51. *IAR,* 59-60, P. 48ff.

52. Chakrabarty, M. 1967, *Op.cit,* p. 106

53. *IAR,* 81-82, P. 34

54. De, Sudhin, 1989, 'The Osteo Keratic Culture and Faunal Exploitation in Protohistoric West Bengal' Paper read in the National Seminar on Archaeology in Eastern India.

Dasgupta, P.C. 1967, *Pragaitihashik Susunia* (Bengali), Calcutta.

55. Pal. A.C. 1996, *Prehistory of the Susunia Hill Complex and the Subarnarekha Valley, DAWB.*

56. Chakrabarti, Dilip K. 1990, *Op.cit,* P. 149.

57. *IAR,* 68- 69, P. 57

58. Chakrabarti Dilip K, *etal,* 1983, 'Lithic Industries of Purulia', *ME,* VII P. pp. 131-141.

59. Chakrabarti, Dilip K. 1990, *Op.cit,* P. 122.

60. Roy, Reba, 1987, *Ancient Settlement Pattern of Eastern India,* Calcutta. P. 273.
IAR, 63-64, P. 59

61. *IAR,* 84-85, P. 97

62. Bhattacharya, M. 1983, *Op.cit,* pp. 124-132.

63. Chakrabarti, Dilip K. 1990, *Op. cit* p. 125.

64. *IAR,* 73-74, P. 32

65. Sant, U. 1991, *Neolithic Settlement Pattern of North Eastern and Northern India,* Delhi, P. 50

66. *ibid,* P. 52

67. *IAR,* 72-73, P. 36

68. *IAR,* 82-83, P. 53

69. Bhattacharya, M. 1983, On the Recent Discovery of Ground and Polished Stone Industry of South West Bengal, in D. C. Sircar felicitation vol. pp. 103-107

70. *IAR,* 75-76, P. 58

71. *IAR,* 71-72, P. 81

72. Chakrabarti, Dilip K. 1990, *Op. cit* P. 179.

73. *IAR,* 75-76. P. 57

74. Chakrabarty, M. 1967, *Op.cit,* P. 106

75. *ibid,* P 103

76. *ibid,* P.104

77. *IAR,* 60-61, P. 68

78. Chakrabarty, M. 1967, *Op.cit,* P. 107
IAR, 61-62, P. 60

79. *IAR,* 66 -67, P. 7

80. Sen,D. 1969, 'Profile of a Preliterate Culture in the Sanjai valley, Singhbhum', *JIAS* IV, pp. 17-29.

81. Sen, D. 1975, 'Dungi, Profile of a Neolithic Site', *JIAS,* X, pp. 159-164.

82. Sen, D. 1969, *Op.cit,* P. 23

83. Sen, D. 1975, *Op.cit,* P. 161

84. *IAR* 62-63 P.6

85. Sen. D. 1950, A Celt Site in Singbhum', *MI,* XXX, P.I

86. Sen, D. 1987, *Op.cit.* P. 211.

Sen, D. 1960, 'Preliminery Note on Neolithic Typology of Chakradharpur' *PISC*, (28th session), Pt. Ill, P.214

87. Roy, G.S. 1954, 'Microlithic Industry of Bongara', *MI,* XXXIV, pp. 16-19.

88. *IAR, 61-62,* P.9

89. *IAR, 76-77,* P.9

90. *IAR, 59-60,* P.4

91. *IAR, 59-60,* P.14

92. *IAR, 59-60,* P.4

93. *IAR, 72-73,* P.8

94. Chakrabarti, Dilip K. 1990, *Op.cit* P. 64

95. Dasgupta, H.C. 1918, On a Peculiar Polished Hammerstone from Singhbhum, Chhotanagpur, India', *IA,* XXIII, pp. 135-136.

96. *IAR,* 85-86, P.12

97. *ibid,* P. 13

98. *ibid,* P. 14

99. *ibid, P.*14

100. *The Sunday Statesman*, 1998, 4th Aug. P. 5

101. Wood Mason, J. 1888, 'Notes on Some Objects from a Neolithic Settlement Recently Discovered by Mr. W.H.P Driver at Ranchi in the Chhotanagpur District', *JASB,* pp. 387-396.

102. Wood Mason, J. 1889, 'Notice of a Neolithic Celt from Jashpur in the Chhotanagpur District', *JASB.,* P. 254.

103. Ball, V. 1878, 'On Stone Implement from Chhotangpur', *JASB,* P.118.

104. Roy. S. C. 1915, 'Stone Tools from Ranchi', *JBORS,* pp. 111-118.

105 Ghosh, A. 1961, 'Stone Tools from Midnapur', *SC,* XXVII, P. 67ff

106. Bodding, P.O. 1901, 'Ancient Stone Implements in the Santal Pagranas', *JASB,* pp. 17-22.

107. Allichin, F.R. 1962, 'The Neolithic Stone Industry of the Santal Pargana', *BSOAS,* XXV (ii) pp. 306-330

108. *An. Rep. ASI*, 1923-24, pp. 100-101

109. *Ibid,* pp 100-101

110. Brahmachari, U.N, S.C. Brahmachari, 1926, 'Neolithic Stone Implement Found at Jamalpur (Monghyr)', *JASB,* XXII (N.S) P. 134ff.

111. *IAR,* 82-83, P. 26

112. Chakrabarti, Dilip. K. 1990, *Op.cit,* P. 22

113. *ibid,* P. 34

114. *The Statesman*, 24th Nov. 1996, P.8

115. Bose, N.K. 1928, Age of Boulder Conglomerate at Kuliana, *PISC,* P. 21

116. *An. Rep. ASI*, 1923-24, P. 100

Roy, G. 1960, 'Prehistory of Orissa', *MI,* XXXX (III), pp. 14-16.

117. Sahoo, D. 2000, 'Prehistory of Darpankhas (District Jajpur) Coastal Orissa' in P. Mohanty, K. K. Basa (eds) *Archeaology of Orissa.* Vol. 1, pp. 173-200.

118. Behera, P. K. 1992, 'Pebble Tool Component of Bonaigarh Neolithic Complex, Orissa', *ME,* XVII, pp. 57-63

119. Roy, G.S, 1960, *Op.cit*, P16

120. *IAR,* 84-85, P. 60

121 *Ibid,*p.61

122. *IAR,* 83-84, P. 52

123. Kar, S.K. 2000, 'Gopalpur a Neo-Chalcolithic Site of Coastal Orissa; in K.K. Basa and P. Mohanty (eds) *Archaeology of Orissa,* pp. 368-391.

124. *IAR,* 58-59, P. 38

125. *IAR,* 58-59, P. 36

126. *IAR,* 84-85, P. 57

127. *IAR,* 58-59, P. 36

128. *IAR,* 84-85, P. 57

129. Singh, P.K. 1972, Cultural Succession of Prehistoric Culture in Central Orissa in K.L. Bhowmick (ed) *Culture of the Past* P. 95 ff.

130. *IAR,* 83-84, P. 39

131. *IAR,* 58-59, P. 73

132. Mohanty, P. 1986-87 'Stone Age Research In Keonjhar District Orissa, *Puratattva,* XVI, P. 54.

133. Sharma, Y. D., 1952, 'A New Stone Implement from Orissa', *MI,* XXXII (I), P. 14.

134. Basa, K.K. T K Das *etal*, 2000, 'Neolithic Culture of Pallahara, Central Orissa', in P. Mohanty (ed) *Archeology of Orissa,* Vol. 1, pp. 325-355.

135. *IAR,* 56-57, pp. 30-31

136. *IAR,* 82-83, P. 64

137. *IAR,* 1982-83, P. 34

138. Assam Research Society, 1984, *A Source Book of Archaeology of Assam and Other States of Northeastern India,* P. 12

139. *Ibid,* P. 18

140. Sankalia, H. D, 1974, *Op.cit,* P. 291

141. *Ibid,* P. 291

142. *Ibid,* P. 291

143. Goswami, M.C, A. C. Bhagabati, 1959, 'Preliminery Report on a Collection of Neolithic Tools from Western Assam', *MI,* XXXIX, P. 312 ff.

144. *ibid,* P. 315

145. Sankalia, H.D. 1974, *Op.cit,* P. 291

146. Rao, S.N. 1973, *Op.cit,* pp. 1-9

147. Sankalia, H.D. 1974, *Op.cit,* P. 292.

148. Medlicott, H.B. 1875, 'A Stone Hatchet from Dibrugarh', *PASB,* P. 159.

149. Mills, J.P, J.H. Hutton 1929, 'Ancient Monolith of North Cachhar', *JPASB,* XXV (N.S), P. 295 ff.

150. Brown, J.C. 1914, 'Grooved Stone Hammer from Assam and Similar form in Eastern Asia', *JPASB,* IX (N.S), pp. 107-109

151. Dasgupta, H.C. 1913, 'On Two Shouldered Implement from Assam', *JASB,* IX(N,S), pp. 291-93

152. Bhagabati M C, A.C Goswami, 1960, 'A Typological Study of Shouldered Celt from Rengchangiri', *JUG,* X pp. 105-12.

153. Ray Reba, 1987, *Op cit.* P.289.

154. Dani, A. H. 1960, *Prehistory and Protohistory of Eastern India,* Calcutta pp. 62-70.

155. Sharma, H.C. 1976 , 'Stone Age Sites in Garo Hills Meghalaya', *Bull. Dept Anthropology,* Gauhati V pp. 82-93.

156. *IAR,* 66-67, P. 4

157. Assam Research Society, 1984, *A Source Book of Archaeology of Assam and Other States of North-East India,* P. 53.

158. *IAR,*, 66-67, pp. 3-5

159. *Ibid,* pp. 3-5

160. *IAR, 69-70,* P. 2

161. Assam Research Society, 1984, *Source Book of the Archaeology of Assam and Other States of North-East India* P. 52

162. *Ibid*, P. 59.

163. Assam Research Society, 1984, *Op.cit,* P. 56

164. *Ibid,* P. 65

165. Sankalia, H. D. 1974, *Op.cit*

166. Godwin Austen, H.H. 1875, A Celt Found at Khasi Hills Shillong', *PASB,* P. 158

167. *IAR* 92-93 P. 100

168. Lubbock, 1867, 'Stone Tools of Assam', *Athenaeum,* 22nd June

169. *IAR,* 69-70, P. 30

170. Bopardikar, B P 1972. Prehistoric Investigation of Daphabum in S B Deo (ed) Archaeological Congress Seminar Paper. pp 1-8

171. Assam Research Society, 1984, p. 59.

172. Ashraf, A.A. 1990, *Prehistoric Arunachal: A Study on Prehistory and Ethnoarchaeology of Kamla Valley,* Directorate of Research, Arunachal.

173. Chatterjee Subid, 1991, *A Comprehensive History of Arunachal Pradesh,* Calcutta;P.28.

174. Sankalia, H.D. 1974, *Op.cit,* P. 293

175. Ashraf, A.A, 1990, *Op.cit,* P 32

176. *IAR,* 69-70,P 2

177. *IAR,* 83-84,P 26

178. Singh, O.K. and T.C Sharma, 1969, 'On the Discovery of Stone Age Relics from Manipur'*JASS,* XII, pp. 36-48.

179. *IAR,* 80-81, P. 12

180. *IAR,* 86-87, P.47

181. Sankalia, H.D. *Op.cit*

182. Singh, O.K. 1980, *Archaeological Research of Manipur,* Govt of Manipur P. 149

183. *ibid,* P. 152

184. Singh, O.K. 1991, *Neolithic Stone Tools of Manipur,* Kakching.

185. *IAR,* 80 -81, P. 38

186. Steel, E. H. 1870, 'Celt Found Among Namsang Nagas', *PASB,* P. 267

187. Peal, S.E. 1872, 'A Model Illustrating the Naga Method of Climbing Trees', *PASB*, pp. 135-138.

188. Hutton, J.H. 1924, Two Celts from the Naga Hills', *Man,* pp. 20-22.

189. *IAR,* 9 1-92, P. 112

190. *ibid,* P.112

191. Sharma, T.C. 1984, 'Prehistoric Culture of North East India', *Indian Society for Prehistoric and Quarternary Studies Seminar Paper.*

192. Ramesh, N.R. 1986, Discovery of Stone Age Tools from Tripura and Its Relevance to the Prehistory of South East Asia, *Geological Society of Malaysia,* II, pp. 289-310.

193. Randhawa, M.S. *A History of Agriculture in India,* N. Delhi, P. 107.

194. Chakrabarti, Dilip.K. 1998, *The Issues in East Indian Archaeology,* Delhi, p.82.

195. –1999 *India: An Archaeological History,* Delhi, p.207.

196. -, 1998, *Op.Cit* P. 104

197. Biswas, M.K. 1980, 'Animal Remains of Chirand', *Bihar Records of ZSI,* LXXVI pp. 115-124.

198. Banerjee, S. 1981, 'Animal Remains from Bharatpur (District- Burdwan West Bengal, India), *Records of ZSI*, LXXIX, pp. 193- 201.

Chapter- 4

Ethnography

From time immemorial India has figured as a prominent place inhabited by early forms of life. Although no fossil evidence of stone age *Homo sapiens* has yet been discovered from either eastern or north-eastern India, the discovery of prehistoric skeletal remains from other parts of the country and earliest forms of stone tools from all over India, including the east and the north-east, has established, beyond doubt, its position among the countries inhabited by human population in the remote past. The subcontinent represents a confluence of diverse languages and cultures. She served as a deep net in which various communities of Asia have been drifted and caught.

People entered here in groups and settled down in different parts. With the lapse of time other groups came and sometimes the later immigrants subjugated the former physically as well as culturally, but even then often the former manages to carry on with some of their cultural, physiological and linguistic identities, losing some of them. Moreover, with the advancement of technological knowledge, people discarded many of their old cultural traditions themselvs and adopted modern devices to make their livlihood more comfortable. In these processes only some specific items survive, may be in a modified form and are handed down in posterity. Such remnant of survival is an important key to the past. As nothing 'better than the existing, living evidence can explain the process by which cultural systems were maintained and modified in the prehistoric past[1]', we shall look into such survivals.

The study of this living community is of cardinal importance to an archaeologist, trying to reconstruct the past. As for judging 'how mankind may have once lived it is also a great help to observe how they are actually found living'[2]. The technology, material culture, economy and even social and religious practices of these folks and primitive societies are valuable guide to suplement the fragmentary archaeological record.

This subcontinent proudly posseses the largest concentration of tribal population of the World. According to census of 2001 the schedule tribe constitutes 8.2% of the whole population of the country[3]. The forested hilly north-eastern India might have been a region of comparative isolation, but other parts of eastern India also allured people with their fertile soils and congenial climate from a remote past.

In search of the past, the tribes of our concerned region will be studied in this chapter, to have more knowledge about the past people. In addition to present ethnographic evidences, past records of these aboriginal people might fit well with the prehistoric evidence. Thus, the whole region under study is a potential area for ethno-archaeological survey. It would be pragmatic to have a general idea about the population distribution of our study area and gather some information about their way of living.

The mountainous north-eastern India, with its two fertile plains divided by Khasi and Jaintia hills, is situated between the eastern Himalayas, southern Tibet, China, Burma and Bengal. This area has received immigrants from all these bordering countries. Thus cultural-linguistic affinities of these tribes with its neighbouring people of East and South-Eeast Asia indicate a common origin. Among the seven sister states of north-eastern India, Mizoram exhibits the largest concentration of tribals (about 94.75%) followed by Nagaland (87.7%), Meghalaya (85.5%), Arunachal Pradesh (63.65%), Manipur(34.7%) Tripura (30.99%) and Assam (12.8%)[4].

In this mountain clad country the splendid tribes like the Garo, Khasi, Naga, Mech, Rabha, Kuki, Chakma, Miri, Hajong, Jaintia, Mizo, Mishmi and others lived in comparative isolation, till modern times, retaining some of their own traditional habits and institutions though proselytized to some extent.

The **Khasis** are a Mon-Khmer-Nicobar groups of Austro-Asiatic language speaking people particularly spread over Assam and Meghalaya. Physically broad headed, short to medium statured dark skinned Khasis are regarded as an offshoot of Mon people of further India and show a very close approximation to the Malaya type. From Khasi grammar Rev. H. Roberts connects the Khasis politically with the Burmese[5]. Geldern was not sure whether the first cousin of the Khasis, the Mons, migrated to their present abode from China or Burma[6], though Risley believes 'the first Indo-Chinese to invade the territory were the ancestors of Mon Khmer whom we find .... surviving as Khasis'.[7] Primarily the Khasis are agriculturists though hunting is still practised. In making of pottery, potter's wheels are still not used. They use a peculiar shouldered iron hoe (*mokhiw*) in ther agricultural field (Plate 32A), which will be discussed in detail in next chapter.

Included in the broad term of Khasi, the **Jaintia** tribe also belongs to Mon-Khmer family. They seem to have been one of the earliest groups of migrants from South-East Asia. The tribe is now distributed mainly in Assam and Meghalaya. They speak in *pnar* language. In comparison to some other north-east Indian tribes, agricultural technique of the Jaintias is far advanced. Besides agriculture, they practice other avocations too *viz*. trade and commerce.

The dark complexioned, short statured, matrilineal **Garo** tribe is found over Meghalaya and Assam. Regarding the origin of this tribe, they believe themselves to be autochthonous, but according to a legend the ancestors of the Garos originally inhabited a province of Tibet. They belong to the Tibeto-Burman linguistic stock. This tribe is essentially cultivator. 'Cultivating the soil', says Major Playfair, 'with him is the beginning and the end of his life's work and the occupation to which he devotes all the energy he possesses'.[8] In farming they use a home made hoe (*gitchi*) (pl 32B), digging stick, etc which are discussed in detail in chapter V. Head hunting was once a favourite passtime of this people, but it was effectively stopped in year 1876.

Earlier called **Mikir**, the tribe now prefers to call themselves either as Arleng or Karbi. According to legend original home of this tribe lies somewhere in South-East Asia. Their present address is the districts of Assam and Nagaland mainly.Their language belongs to the Bodo group of the Tibeto-Burman family. They practice both shifting and settled cultivation, besides hunting and fishing. Their pottery is hand made. They have no idols, temples or shrines, great boulders besides many a natural objects are worshipped.

Fair complexioned, broad headed **Chakmas** are distributed over Assam, Tripura, Mizoram and West Bengal. They speak in *chakma*, an Indo-Aryan form of language.

Regarding the original seat of the Chakmas, there is ample scope to controvert over it. The Chakma chronicle endorsed the place named 'Champak nagar' as their original home. Several places bearing this name are situated in different places like Burma, Bangladesh, Malay etc. It can only be suggested that their original seat was somewhere in the East or in South - East Asia. Traditionally the Chakmas are shifting cultivators. In addition to that, settled cultivation, hunting, gathering are other economic pursuits of the tribe.

Mishmi is a short and sturdy people, with well knit, broad shouldered figure. They are found mostly in Arunachal Pradesh. This tribe is mainly engaged in trading and less to cultivation. They are very rich in their herds. Their language *mishmi* belongs to Tibeto-Burman family.

The **Kukis** are found in the states of Assam, Nagaland, Tripura and Manipur. They are regarded as hunters, though food gathering, fishing and cultivation play an important role in their subsistence economy. Kuki myth dates their migration from China around 200 B.C.E.[9] Though penetration of the Kukis into Assam hill continued upto present time, their distribution in marginal areas and inaccessible forest and mountain tract suggest an early immigration of this tribe into India. Their language has been classed among the Austro-asiatic group of Tibeto-Burman family. This tribe use a Neolithic type axe for tearing out weeds from wet ground, which is discussed in next chapter.

A very ancient legend speaks that the habitat of the **Mizos** were near the thick wall of China, and the search for a healthy home brought them to north-eastern India. They are now settled over Mizoram, Meghalaya and Tripura. The Mizos with their keen sparkling eyes, snub nose and wheatish complexion belong to Kuki-Chin group of tribes. Linguistically they belong to Tibeto-Burman family. Economy of the Mizos completely tied down to agriculture. They also practise fishing with nets as well as by the old method of poisoning the entire water of a pond. Besides that they are professional craftsmen.

Adi is one of the oldest and culturally quite advanced tribe of Arunachal. A feature which characterises the tribe is their love for the tradition, which is meticulously followed since an unknown past. Their past history is very little known. Adi myth and folklore tell that, the ancestors of various groups of Adis lived in Tibet. Bhattacharya recognised the trijunction of India, Burma and China as the original abode of this people[10]. Adis consist of numerous groups. They speak in *bokar* language of Tibeto-Burman family. The tribe practices both shifting and terrace cultivation.

From the Shan country of Burma, through the hills of Chittagong of Bangladesh, the **Riangs** entered in Tripura region.[11] They are short to below medium statured people belonging to Mongolian-Kuki group. Their mother tongue *riang*, belongs to Tibeto-Burman family of language. Agriculture is the mainstay of the Riang but they have to depend on other professions like carpentry, basket making, fishing, hunting, cattle rearing etc. to earn their bread.

The **Nishi** or **Dafla** tribe is mainly settled in Kameng and Subansiri district of Arunachal Pradesh. They are very proud and virile people. They speak in *nishi* which belongs to Tibeto-Burman family of language. Myths hint at eastern corner of the Himalays as their original habitat. Agriculture is the main economic activity of the tribe, besides that livestock rearing has significant role in the socio-religious life of the tribe. In farming they use hoe (*lowdap*), axe (*alengahey*) (fig. 15 A,B), chopper (*da*) etc. alogn with other tools. They are expert hunters also. Only Dafla women ply the craft of pottery making and their pottery is locally known as *takam pachang*.

The **Khamptis** inhabit the plains of Lohit-district of Arunachal Pradesh. Their mother tongue *khampti-shan* belongs to the Siamese Burman family of languages. Traditionally they are cultivators, timber merchants and business men. This tribe hafts their hammer stone in a very rudimentary and simple type of handle made of creeper, which guide us in reconstructing our archaeological tool, are dealt in next chapter.

Tangsa is an umbrella term for a number of communities living in the mountain terrain of Arunachal Pradesh. Each group speaks a separate dialect of the Tangsa language, which belongs to the Tibeto-Burman family of languages. They believe that they originated at Masoi Sinrapum, a place somewhere in the east. The tribe is jhum cultivator. In cultivation a special type of tangged hoe (*jankho*) (Fig.15C) is used, which will be discussed in the next chapter. The tribe is expert in the craft of weaving and basket making. Besides that they also engage themselves in hunting, fishing etc.

Another Mongoloid tribe of Assam region is the **Mech**. They are largely found in West Bengal also. Their language has been classed as belonging to Tibeto-Burman group. The Mechs possess a rich vocabulary. More than half a century

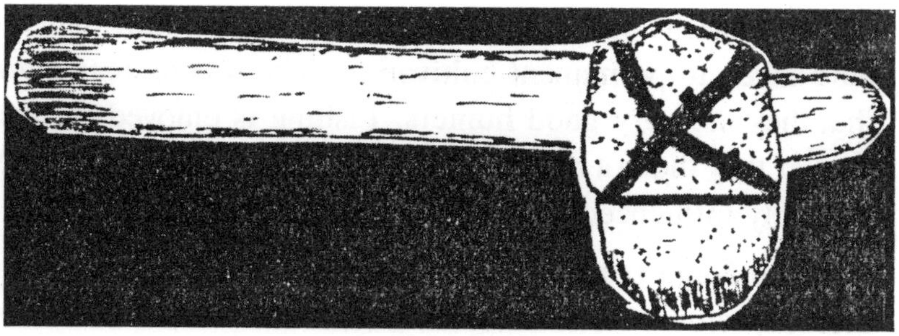

Fig. B

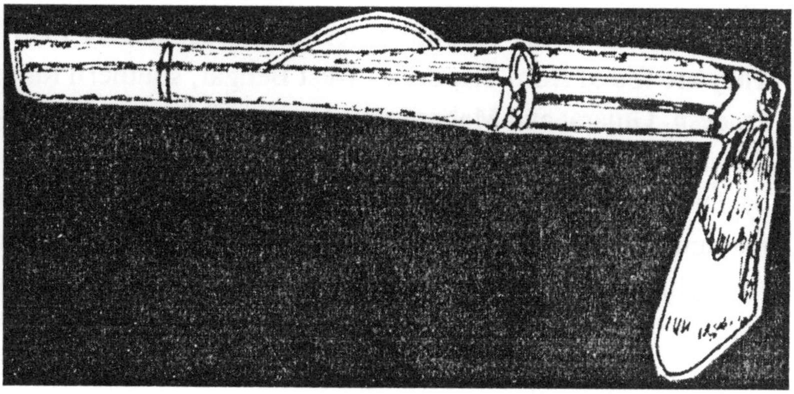

Fig. C

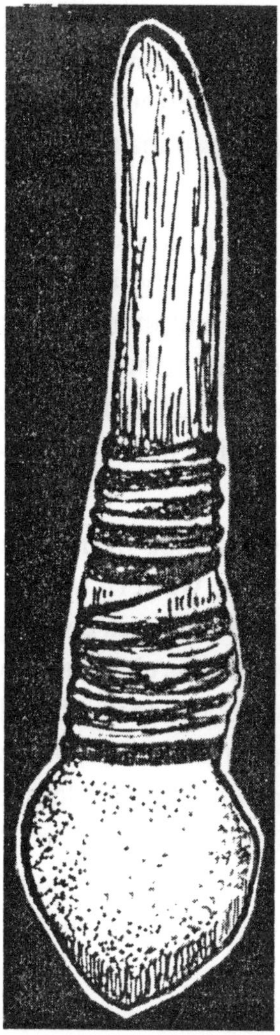

Fig. A

Fig. 15. A. Stone hoe of the Nishi B. Stone axe of the Nishi. C. Stone hoe of the tangsa

ago, the Mechs practised jhum cultivation, but now they are rooted to the soil and have taken to permanent cultivation. The Mechs rear animals and birds. Besides that they are very good hunters. Fishing is enjoyed for sport as well as for food. Earlier they did not believe in idol worship but lately they have started making idols, probably influenced by the Hindu culture.

The **Naga** tribe is predominantly Mongoloid. The stories of their origin tell their movement from different South -East Asian countries. This tribe is divided into different sub groups like Ao, Sema, Konyak, Phom, Lhota and many others. Land, forest and water are three main economic resources of this tribe. All their principal indigenous household utensils are made of clay bamboo or gourd (Plate 31D). Different objects of their material culture shows similarity with Neolithic artifacts which will be detailed in the next chapter.

The **Bhil** is the second largest and most widely distributed tribe over India. They are found from Tripura to West Bengal, southern Rajasthan, Western Madhya Pradesh, Gujarat and Maharastra. Their traditional home constitutes an area covering the hills like Aravalli, Vindhya and Satpura plateau. Physically a Bhil is characterised by dark brown complexion, short to medium stature, well built body, sparse beard, medium to long head etc. Haddon has grouped them as a branch of Pre-Dravidian. According to J.H. Hutton 'the Bhils appear to be a mixed type entailing Caucasian and Australoid elements and sometimes features which rather suggest a touch of Mongolian[12]. They speak in *bhili* language, which belongs to the Indo-Aryan family of language. Bhils are mainly cultivators, besides that their economy also stands on profession like domestication of animal, fishing etc.

Numerically the most dominant tribe of India is the **Gond**. Their homeland extends from the Satpura range down to the Godavari, from Uttarpradesh and north Bihar to Andhra Pradesh and from Maharashtra to Orissa. According to Schedule Caste and Schedule Tribe orders act 1976, the Gonds have more than fifty groups. Gonds deserve special interest for the fact that they are the only Indian tribe to set up a kingdom and ruled over greater part of central India for about four centuries starting from 14th Century C.E. Formerly food gatherer Gonds are now engaged in a variety of occupations like day labour, carpenter, field labour, artisan etc. Earlier reports about the Gonds tell that they were a famous hunting tribe. They have a rich tradition of art and craft. They have strongly built body with features well cut. Risley grouped them among the Dravidians and Dalton among the Pre-Dravidians. Their mother tongue *gondi* is similar to that of Dravidian language.

The **Santal** is one of the main tribes of India, contributing about 9.5% of entire tribal population of this sub-continent. At present they are distributed over West Bengal, Bihar, Jharkhand, Orissa, and Tripura. Their myth gives a complete picture of their migratory route starting from Iran through different countries. They ultimately reached Hazaribagh region of Jharkhand. This hardy, simple hearted tribe is an efficient agriculturist and excellent hunter. Strong co-operative spirit characterises all Santal economic activities. Besides farming and hunting they also work as fisherman, unskilled labourers in coalmines, tea gardens, factories etc. The Santals are a people of very good artistic taste. Their language *santali* is one of the oldest tongue of India.

One of the most well studied tribal community of India is the **Munda**. They claim that they have original settlers right in India. They believe Chhotanagpur as their fatherland where bones of their ancestors lie buried. But according to J.F. Hewitt, Madagascar is the original seat of the Mundas[13]. According to A. C. Haddon the Munda speaking people are a very ancient element of population of India. They speak in *mundari*, one of the major Austro-Asiatic languages of India. They mainly subsist on cultivation while hunting and collecting of forest produce are their subsidiary occupation. Special interest of this tribe lies here in their use of Neolithic type shouldered hoe for agricultural work, which are described in following chapter.

In the *Bhisma-parva* of the *Mahabharata,* name of an aboriginal tribe *Karusha* is mentioned. Grignard[14] suggested these Karushas to be the progenitor of modern **Oraon**. He showed that the Oraons reject their modern name Oraon and does not call himself otherwise than Karukhas. Grignard suggested that 'Karusha' and 'Karukha' are identical. At present Oraon economy is based on hunting, fishing, cattle rearing, agriculture etc.

Bedia is a small community of Bihar and West Bengal. Being a land holding community the primary occupation of the Bedia is agriculture. Occasionally they go for hunting. Bedia men are engaged also in selling of fire wood, collecting of lac and other forest produces besides serving as casual labours. *Panchpargania* which is a form of Indo-Aryan family of language is used for both inter and intra group communication.

Distributed over Bihar, Madhya Pradesh, Orissa and West Bengal, the **Kharia** is a primitive Kolarian tribe. They have derived their name from palanquin. In general the Kharias are of short or below medium statured, long, narrow headed people. Their language *kharia* belongs to Austro-Asiatic family. This non-vegetarian community depends mostly upon the forest resources, besides cultivation.

The **Ho** tribe is mainly found distributed over Bihar. Their dialect closely related to that of *mundari*. Land is their main economic resource which is controlled individually. They primarily subsist on agriculture. They are also engaged in selling of fire wood, collection of forest produces, fishing etc.

Formerly notified as criminal tribe, the **Lodhas** now-a-days subsit on agriculture mainly, besides wood collecting, fishing etc. A high degree of social solidarity is maintained in day-to-day economic activities. According to Bhowmick the Lodhas on an average are of below medium height with a long head, broad face and broad nose. According to the Census of India, the Lodhas originally migrated from central provinces. Mother tongue of this community is akin to Savara, an Austro-asiatic language.

Mech tribe of West Bengal is more advanced than its neighbours. They came from Assam and settled down in West Bengal. Jhum cultivation, thread making, weaving, fishing etc. are their various means of earning livelihood. Their dialect belongs to Assam-Burmese group of Tibeto-Chinese family of language.

In mediaeval Sanskrit literature, the barbars of the Vindhya hills are called in the Purana as **Savara**. The Savara is a very ancient tribe of India. It is believed that the Savaras were definitely the maker of Stone Age culture[15] but any direct proof can not be traced. They use shouldered adze for agricultural purposes. Cunninghum suggested that the Savaras at one time covered all the different divisions of Kol, Juang and other Kolarian tribes. Savara economy is mainly supported by forest produce. Presently they have adopted the profession of snake charming.

The term **Birhor** means jungle man. Forbes states that Birhor is an aboriginal tribe of very low type and are doubtless mundari. Birhor dialect belongs to the Austric family of Austro-Asiatic language. Birhor settlement is found over Bengal, Bihar, Jharkhand and Orissa. This nomadic tribe subsists on jungle products mainly.

The **Juang** tribe is medium statured and black to brown skin coloured community. On the basis of their linguistic affinities the Juang are grouped as mundari. Datlon says Juangs are in habits and customs the most primitive people and considered them as survival of 'the stone age in situ'[16]. Land and forest provide them main resources of livelihood, besides a large section are engaged as agricultural labour, hunter, fisherman, basket maker etc. They use digging stick and hoe for tilling the soil

So, it could be said that eastern and north-eastern India have a long history about the study of different population groups, mostly carried out by the British officers, deputed in this area. We have seen the Tibeto Burman speaking community predominate the aboriginal population of north eastern India and most of them possess a legend of immigration from East or South East Asia. According to Van Driem homeland of this Tibeto Burman speaker is Szechwan Yunan. This group had started a southerly migration from its original homeland around 7000 BCE.[17] It is also well known that scholars like Risley, Dalton, Hunter *et al* not only studied the material culture of different tribal communities of different corners of eastern and north-eastern India, they have tried to highlight the potentialities of this population to understand the cultural milieu of this region. It is very difficult to locate all these people in the context of our Neolithic culture and economy because by and large the tribal groups which were mainly hunter-gatherer move in and around different landscape of eastern and north-eastern India, in search of pasture and agricultural land, whereas there are a sizeable population tends to show the change of economy and behavioral pattern. Thus only a tentative picture of hunter-gatherer to village farming economic development could be realised from the survey of some population of this area. Therefore the pastoralist class, the agricultural class, the potter class etc. are the basis of our knowledge for corroborative study. Though it is very difficult to record all these types as the continuation of Neolithic folk but it may not be unwise if we render our suggestion with older days including Neolithic period. Therefore our summary about population which has presented here may help us to understand Neolithic character as a basis of ethnoarchaeological study.

References:

1. Stiles, Daniel, 1977, 'Ethno Archaeology : A Discussion of Methods and Application', *Man* (N.S), XII, pp. 87-103.
2. Nagar, M. 1975, 'Role of Ethnographic Evidence in Reconstruction of Archaeological Data', *EA,* Vol XXVIII (I), P.13.
3. Census of India, 2001, Series IV.
4. *ibid*, 1907.
5. Gurdon,RR.T, *The Khasis,* London, P. 10.

6. *ibid*, P. 12.
7. Risley, H.H. 1915, *The People of India,* Calcutta., P. 116.
8. Playfair, A, 1909, *The Garos,* London, P. 11.
9. Gangte, T.S. 1993, *Kukis of Manipur; A* Historical Analysis. N.Delhi P. 41.
10. Bhattacharya, T.K, 1979, *The Adis - Origin and Migration.* Calcutta.
11. Chatterjee, Suniti Kumar, 1951, 'Kirata - Jana-Kriti : The Indo Mongoloids : Their Contribution to the History and Culture of India' *RASB,* Calcutta. P.9.
12. Hutton, J.H. 1931, 'Ethnographic Notes by Various Hands',*COI,* vol 1 pt III.
13. Hewitt, J.F. 1972, *The Ruling Race of Prehistoric Times in India, South Western Asia and Southern Europe,* Delhi, P.62.
14. Grignard F.A. 1909, 'The Oraons and Mundas from the Time of their Settlement in India'. *Anthropos,* Tom IV, heft 1, pp. 1-47.
15. Roy, H.M. 1986, *Savara, The Snake Charmer,* Calcutta, P.12.
16. Dalton, Colonel, 1872, *Descriptive Ethnology of Bengal* Calcutta, P. 225.
17. Driem G V 1998 Neolithic Correlates of Ancient Tibeto Burman Migration, In Blench and Spriggs (eds) *Archaeology and Language Ii: Archaeological Data and Linguistic Hypothese.*

    Chowdhury, J.N. 1991, 'Prehistoric and Early Tribal Migrations in North-East India, in J.P. Singh and G. Sengupta(eds) *Archaeology of North-Eastern India,* N.Delhi, P. 104.

Chapter 5

Ethnoarchaeological Study

The sub continent of India is divided almost equally into two by the tropic of cancer. That is why the northern part of the country is dominated by sub-tropical and temperate climate. In this type of climate organic materials decay very fast and thus do not favour preservation of human remains also. So, in the understanding of Indian culture, archaeologists are miserably ignorant about its chief resource *i.e.* the prehistoric man, rather profuse dressed stones are left behind to ponder over and speculate about its maker. But the ultimate aim of an archaeologist is not to dig up stone tools, rather to be engaged in the study of man, because culture is not the sum total of tools or potteries, but an inter-related system of its components. In the study of early human culture, the archaeologists want to get much nearer to early man, to see him as an individual, as a member of a society. The concept of past society is something which can not be put to spade. Nobody could excavate a political system, a set of religious beliefs or social customs or climatic cycle and such set of behavioural pattern that together forms the culture of a society, what an archaeologist would ideally likes to reconstruct.

The solitary memorial of the Stone Age, *i.e.* the imperishable stone objects could not last the ravages of time with few other materials, which, except in very rare instances, only partly survive in the long run. These reduced, decayed and patinated evidences are the only equipment available to a prehistorian, in the study of the past.Thus, archaeological data have their limitations. The only hard evidence described above also could not speak enough about their identity, function or functional values by themselves. Unless the bits and pieces of the same become alive to an archaeologist, it will remain virtually unutilized in archaeological reconstruction.

So, in the attempt of understanding the culture of prehistoric man, there is no other alternative choice before an archaeologist than to exploit every possible resource to make this dumb tool speak. Thus the subject gradually emerged into an interpretative discipline co-operating and collaborating with other fields of study to give a more comprehensive view of the past. The stone, bone or pottery pieces which are otherwise mute begin to tell the story only when studied by a prehistorian. The soil on which they lie could only be analysed by a geologist. The environment that surrounds it could only be examined by a geographer. The man who manufactured them could be investigated by an anthropologist only and so on. Thus it is evident that no one branch of natural science would have been capable of dealing with the problems of archaeology. In this way step by step, piece by piece, the history of the bygone people could be built if such inter disciplinary approaches are adopted and corroborated.

India has always been regarded as a land with sound tradition of her own, in respect of, cultural or social awareness as well as in other aspects of life, may it be subsistence or craftsmanship. Here in India, the past has never been divorced from its present, rather it has been matured without obliterating and absorbing that, which has gone before. As Kosambi observes, 'perhaps nowhere in the world, can such parallels be found more readily than in India'[1]. Many aboriginal tribal culture here, are recongnisable which stem from the prehistoric period. So, when we deal with Indian archaeology no other discipline than ethnography emerges as most suitable for a proper understanding of the value and significance of the prehistoric artifacts.

'Ethnography' is a record of observation of contemporary behaviour. The discipline includes the study of present tribal population. Many ethnographic informations suggest tribals are certainly the cultural descendants of prehistoric people, which helps to gain insight into the prehistoric life patterns. The technology, material culture, economy and even social and religious practices of these folks and primitive societies can be a very valuable guide in interpretation and reconstruction of the fragmentary and somewhat unintelligible archaeological evidence.

It should be remembered that no society, howsoever primitive, is expected to remain static using stone objects over a length of time and it is impossible for a community to live exactly in the same way as it was in the Stone Age. Ethnographic data could be useful only in furnishing the functional possibilities of excavated features and artifacts. Thus it could be said that making observation in contemporary communities would be utilised to secure 'as many as varied interpretative hypothesis as possible to help us understanding archaeological remains'[2]. It could facilitate the development and refinement of insight into the past behaviour when strong similarities could be established between the two. As the society is changing in a rapid pace, the ephemenal nature of ethnographic data is dying day by day and can never be retrieved.

In the attempt of understanding primitive culture, we have drawn upon two main sources - archaeological and ethnographical. The combination of archaeology *i.e.* the study of the material remains of the past and ethnography *i.e.* the study of the culture of the present, or what we may call ethno-archaeology should be a welcome approach to the study of archaeological remains. It is a study of living communities from the point of view of archaeological evidence. In Stainslawaki's word 'Ethnoarchaeology is the direct observation field study of the form, manufacture, distribution, meaning and use of artifacts and their institutional setting and social unit correlates among living, non-industrial people for the purpose of constructing better explanatory needs, to aid archaeological analogy and inference'[3]. Ethnoarchaeology could be directed at any facet of culture. It is in a sense a technique for the recovery of information and applicable so long as the human activities leave a material residue. Chang[4] viewed ethnoarchaeological study as the reconstruction of an old play that is being staged at another theatre but in same stage setting. It provides substantive empirical data which supplement available ethnographic data and this data are frequently utilised in the evaluation of hypothesis derived initially from interpretation of archaeological data or ethnographic record. So, inspite of its criticism, it could be said that ethnoarchaeology provides a reasonable degree of hypothetical probability than mere speculation and romanticising about the past.

In ethnoarchacology, it is attempted to reach the past via the knowledge about the present. But in doing so, a caution should be voiced. If we imagine a two-sided equation, one side of which is placed the contemporary behaviour observed among a contemporary primitive folk and the other by observation of archaeological materials and if more than an apparent similarity is noticed, then only the model could be considered to be suitable for ethnoarchaeological study.

It should not be assumed that all the past behaviours have analogues available for observation today and if hiatus of time between these two becomes very wide, then an ethnoarchaeologist should remain very cautious to draw valid inference. Again a researcher of this discipline might not find all his relevant information in an ethnographic report, since an ethnographer is concerned more about social aspects of people which are not directly represented in archaeological remains.

As Binford opines, 'Our understanding of the past is not simply a matter of interpreting the archaeological record by analogy to living societies as has been commonly asserted. Our knowledge is sound to the degree that we can verify our postulates scientifically......Scientific verification for archaeologists is the same as for other scientists: it involves testing hypothesis systematically'[5]. So, it would be the best to make analogy between two such models, where a more or less similar economic level and ecological backdrop of limited time gap prevails. Even after that, every source necessitates careful sifting before its incorporation into the study. Ethnographic study would be sought here only to obtain hints to understand the essential basis and character of the culture of a region. According to Sinopoli 'ethno archaeological studies in South Asia do not provide a blueprint of the past.'[6] She further added that demonstration of link between past cultures with present day hunter-gatherer is not theoretically necessary and nor she suspects it, as justified.

Study of analogy derived from present observation to aid interpretation of the past event was utilised in Indian archaeology quite long ago. John Marshall attempted to avail existing evidences quite extensively to interpret excavated materials of Mohenjodaro. Mackay largely depended on modern ethnographic analogies to interprete the excavated evidences of Mohenjodaro and Chanhudaro[7]. Cunnighum also stressed on ancient forms of objects in use in countryside to identify excavated objects. So it could be said that though it seems to be a new born discipline, some of the South Asian archaeologists used ethonoarchaeological approaches, long before the term came to be coined in the western World.

V.G. Childe[8] observed that it would be impossible to recreate past knowledge on its own terms. He further entrusted the job to historian to locate the past in its proper social and chronological context in those thoughts that are still living and active in our own culture. This assertion is very much relevant in eastern and north-eastern Indian archaeological perspective. This part of the sub continent could be designated as a living human museum where people have left traces of their former presence from time immemorial. The region has accommodated both a varied tribal population, who still subsist on some primitive economic pursuit like hunting, gathering and other distinctive traits as well as a rich prehistoric treasure in every nook and corner of the hill tops, banks of streams or simply scattered in open air sites. These contemporary, non-literate primitive people might be recognised as the representative if not exactly, of the early tool maker, among whom vestiges of early economic adaptation could be found. Prehistoric artifacts can be better understood if they are studied by comparing them with those of the primitive people of today. As Haimendorf observed that though 'Palaeolithic and Neolithic stone industry has become obsolete, the ways of life and the economic pattern of both the older and younger stone ages still persist in aboriginal India'[9]. Some particular traditions are very well adapted to a regional environment and deeply rooted in the cultural psyche, in the myths and symbols and permeate in many aspects of life. Some elements of the prehistoric people might have preserved unintentionally among these aborigines, which have long been abandoned in civilized societies. It is in their day to day practice where these tribal people show remarkable closeness to prehistoric tradition.

Thus materials are all around our region, for study. In this chapter we shall look into those ethnographic reports having some bearing in archaeological reconstruction. By making effective utilisation of this isolated description together, we shall try to unravel the multifarious facets of the culture.

In north-eastern India, most of the Neolithic sites except Selbalgiri and Riat-turen, have no pre Neolithic phase. In Selbalgiri though a microlithic phase precede the Neolithic but transition from Microlithic phase to Neolithic phase could not be established[10]. So, it seems that Neolithic culture of north-eastern India might have a preceeding phase elsewhere; it is not indigenous to this region. Typological affinity of tools as well as pottery between East and South-East Asian with north-east Indian Neolithic, strongly supports a East and South-East Asiatic influence on Neolithic culture of north-east India, further.

Ethnographically too, the mythology of many present day tribals tells about different East and South-East Asian countries as their original homeland. Linguistically as wells as physically too they are akin to these people. So, it could be conjectured on ethnoarchaeological ground that Neolithic culture of north-eastern India is not indigenous [11].

In Neolithic East and South-East Asia, agriculture was practised in *jhum* method. This mixed crop farming method is a continuous process round the year in a single plot. If East and South-East Asian influence on north-east Indian Neolithic traits is accepted, then there is every chance of borrowing of similar agricultural technique *i.e.*, *jhum* in a natural way. In north-eastern India, this method of raising crop is widely practised today.

The Tangsa is a tribe of Arunachal Pradesh. These people are well known for their *jhum* or slash-and-burn cultivation. In clearing their agricultural field, in weeding and scraping, the tribe uses a hoe, locally known as *jankho*[12](Fig. 15c). This hoe has certain distinct features from the socketted and tanged hoe. The blade of *jankho* is thinner. Besides that, it is not so handy and effective. In hafting, the butt end of the blade is ingenuously inserted and tactfully tied into a piece of folded cane. The bended ends of the cane which ultimately fold into one unit are fastened together by another cane strip to form the handle.

In archaeological context, from the same area at Shallong in Tirap district, Neolithic celts have been reported[13]. So, if these two features are combined *i.e.*, application of primitive nature of hafting of the iron hoe in presence of modern handy and effective hafting practice and East and South-East Asian Neolithic type of farming all over north-eastern India today, it could be inferred that in lithic period stone blades might have been hafted in this way and north eastern Indian neolithians, at least, adopted slash-and-burn method of cultivation.

Presence of charcoal from Lahuradewa, suggest slash and burn agriculture. Sanai Tal, Rai Bareli also give evidence of slash and burn agriculture.

In *jhum* or slash-and-burn type of agriculture a plot is used for some years and then left fallow for few years to regain its fertility. So, in this type of agricultural practice farmers have to shift their plot from time to time. Thus it is quite unnecessary for them to raise permanent settlement structures. Thus *jhum* farmers' dwellings are mostly made up of perishable materials like leaf, clay, bamboo, wood, straw etc.

In archaeological context, the non-yield of any settlement evidence from north east India perhaps supports raising of similar temporary settlement structure. These perishable raw materials of such short duration settlements also produce very thin deposit in a stratum.

So in archaeological context the thin Neolithic deposits of north-eastern Neolithic sites and ethnographical support of use of perishable settlement materials by present day *jhum* cultivators together proves, in another way, the practice of *jhum* type of cultivation by north-eastern Neolithic folks. This postulate is further strengthened by East and South-East Asian affinities.

In present context many a tribe raise their dwellings in hill tops, hill slopes as well as near water sources. In Neolithic period also settlements have been discovered mostly in highlands or on the river banks. So, similar attitude in both past and present on the matter of selecting settlement sites unmistakably underlines that Neolithic folks were dependent on riverine creatures along with incipient farming as well as natural food gathering. Selection of high land reflects their precautionary approach to avoid flood attack.

In north eastern India most of the Neolithic sites are located at hill tops. In hills, stretches of land for cultivation is not available neither river based wet cultivation is possible. Thus slash and burn method of cultivation is a common practise there at present. Similar system of cultivation could naturally be envisaged in hilly locations of Neolithic period.

In the subsistence pattern of present day north-eastern Indian tribals, domestication played a secondary role. Cattle, pig, goat are kept untied and allowed to graze in surroundings after utilised for food, religious rites etc[14]. This type of semi domestication may justify the non-yield of evidence of domestic fauna in north east Indian Neolithic context. From Mahadaha in Belan valley, similar system might have been in vogue, absence of stakes suggest domestic cattle were not kept in captivity in cattle pen.

Ethnographic records suggest that earthen wares are not very widely used by present day tribals of north-eastern as well as eastern India. Most of their household objects are made up of wood, bamboo, leaf made utensils (Plate 31D). In fetching water from streams or water-falls the Garos use bamboo tubes as pipes[15]. Similar bamboo tubes are also used by the Mikirs, Nagas to store their water. Besides that, joints of bamboo tubes serve the purpose of ornament container too[16]. Bottle gourd is often cultivated by different tribes like the Garo, the Juang, etc. for carrying water (Plate 31B)[17]. The Nagas use mithun's horn as drinking vessels. For cooking rice, even newly cut bamboo nodes are often used.

Archaeologically too, an overall review of entire collection of northeastern Indian Neolithic complex reveals that the percentage of yielded pottery is not very significant here. This may be accounted for the same reason *i.e.*, the neolithian might had depended on wooden, unburnt earthen, cane or leaf made utensils like that of many present day tribals of this area. Thin archaeological deposit is also a proof of use of perishable materials for daily use.

Like present day tribal pottery, the Neolithic pottery was also made by coil method and hand modelling. From archaeological site of Parsi-Parlo, Taba and Vigo of Arunachal Pradesh, four classes of pottery have been encountered[18]. Among them the stamped coarse red ware is found with polished axe and waisted axe and has been assigned a late Neolithic date on the basis of typo-technological character similar to one witnessed in China. In present ethnographic context if we consider the Nishi pottery or *takam pachang*[19], it is found that in all its morphological aspect, the pottery remained absolutely unchanged from prehistoric times. This handmade pottery is generally red in colour and decorated with irregular stamped pattern. This very stamped coarse red ware is also encountered in Neolithic as well as historical phases.

Another Characteristic Neolithic pottery of north-eastern India, cord impressed pottery is still manufactured by present day Oinam tribe of Manipur(Hazarika Manjil 2006 Neolithic Culture of North East India: A Recent Perspective on the Origin of Pottery and Agriculture, *Ancient Asia* Vol 1 pp 25-43).

The Sema Nagas fabricate their pottery by hand alone[20] (Plate 31C). After shaping, the pot is fired in an open fire with dried grass and wood. The pottery thus produced is generally coarse and rough. The Neolithic pottery found from Daojali Hading excavation are in all the cases show evidence of imperfect firing. It is evidenced from the dark core seen in almost all the sherds. So, it can be deduced from the ethnographic evidence that the Neolithic folk of Daojali Hading might have also followed the similar open firing method in manufacturing pottery.

Besides these features of a typical Neolithic avocations *i.e.* agriculture, domestication and pottery making, other subsistence economies could be reconstructed with the help of combination of both archaeological and ethnographic evidences. These evidences suggest that inspite of invention of agriculture, Neolithic people depended on earlier economic pursuits like hunting, fishing and food gathering to a great extent. As they were still under the control of nature, farming might have been a seasonal activity only and in lean seasons they had to resort to other substitutes.

Hunting economy could be undoubtedly attested by the use of bow, arrow, spear, etc. by almost all the tribes, even today (Plate 27B,D). Archaeologically, bone spear from Golbai Sasan (Plate 27A), terracotta ball, sling ball from Chirand and Maner (Plate 27 C) suggest their use in pellet bow for hunting.

From Neolithic layer of Chirand, Golbai Sasan, bone spear head have been encountered. Similar spear head of iron are used for fishing by the present day fisherman of north Bihar. Besides using spear head, many primitive tribes of present are found to fish in the pond by poisoning the entire water of the reservoir (Plate 29 B). Further more discoveries of net sinkers, bodkins (Plate 29A) from Chirand, Napachik, Teesta river site etc. also proves fishing activity in Neolithic eastern and north-eastern India. In Chirand, large amount of fish scale and bones found at Neolithic level also suggest fishing was an important pursuit.

In India few primitive tribes are still remained in food gathering stage (forest hunting). They subsist mainly on gathering of various forest produces. For instance in remote areas of Bihar, the seeds of Setaria grass *(Setaria cf gluca)* are found to be gathered by different tribes[22]. This very wild plant with many others have also been found from Neolithic cultural level of Senuwar[23], which indicates towards this vocational practice.

The Khasi, whom we discussed in brief in the Ethnography chapter, constitutes an agricultural tribe. Originally, they represent a shouldered stone hoe culture. In agricultural field they use a peculiar shaped iron spade called *mokhiw* (Plate 32A)[24]. This spade has two shoulders, far projected from its body and according to Gurdon it recalls the shouldered stone celt of north-east Indian Neolithic type. This celt is different in shape from the Neolithic stone celts found from other parts of India (except that of Chhotanagpur plateau), while it shows, similarity in nomenclature with the Burmese celt. Considering these two cases together, it seems that there might be some ground to believe that the Khasis were somehow related with the Neolithic folk of Burma and Chhotanagpur plateau.

The Naga tribe consists of several groups *viz*. Sema Naga, Ao Naga, Kacha Naga, Yimtsung Naga etc. It has already been described that the Nagas also form an agricultural tribe. Various agricultural implements of this tribe provide ample clue in our study.

In digging, the Sema Nagas use a hoe with crooked handle made from a forked bough. It comes in two sizes—the larger one *Pushyekupu* is used for digging up new land and the smaller *hongokupu*[11] is only used in sowing seeds. A variety of this hoe is also found from Yachumi country and is locally called as *tafuchi*[25]. The blade of this type of hoe is nothing but broad blade with shoulders.

The Nagas of Ledo use an iron bladed hoe. The tanged blade of this hoe is presumed by Balfour[26] to have been a derivation of tanged and shouldered type of Neolithic stone adze.

Another division of Naga tribe, the Yimtsung Nagas also use similar little curved shoulder headed adze of Neolithic type[27]. It is generally used in sweet potato patches.

In Santal Pargana many Stone Age relics are encountered, mostly near the settlement of the Mons and Mundas[28]. Among them, shoulder headed celt is mentionable. Scholars are of opinion that these stone implements were originally made by these Mon and Munda peoples.

The Savara is a very ancient tribe of India. Indian scriptures provide ample reference showing their settlement in India long before the immigration of the Aryans. In 1931 census report they have been mentioned as user of shouldered adze, simple bamboo, arrow etc[29]. In the same report[30]Hutton wrote about Oraons, who are also found to use shouldered hoe, not unlike that of the Khasis.

From these evidences it could be safely suggested that the shouldered celt continued its use since the time of the Neolithic period down to the present days. It is further noted that most of these present day tribes use this shouldered celt in their agricultural fields. So, there is room to claim that in north-eastern India, agriculture was known from Neolithic times.

The Garo tribe is generally found in isolated corner of north-eastern India. The livelihood of the tribe is very primitive. The farming Garo use a home made hoe called *gitchi*[31] (Pl 32B) in scraping ground and removing weeds. It consists of a small iron blade set into a piece of bamboo root which serves the purpose of a handle. It is believed that the implement is of indigenous origin. It is also used in the plains of north-eastern India. The shape and size of the blade of the hoe is a perfect replica of the ground stone celt found from Garo hill Neolithic complex.These celts were specially meant for cutting down trees in preparing agricultural land.

In agricultural field the Garos use primeval equipment *i.e.* digging stick or *matham*[32]. It is made from a suitable branch of a tree. One end of it is made pointed to go into the ground smoothly. It is used for planting seeds. Digging stick is known to be used also by the Juang of Keonjhar district, Orissa[33]. In remote past this primitive tribe observed very rudimentary method of cultivation, which consisted of scratching of the earth by thorny branches of tree and broadcasting of seeds over it. Later on digging sticks and hoes were used in tilling the soil. It may be suggested that in primitive agricultural field cereals were dibbled by such digging stick of wood or bone but could not survive for archaeologists in the long run.

It may further be mentioned that ring stones are found from most of the Neolithic sites of eastern India. K.P. Oakley corroborated the association of ring stone with agriculture. Such ring stones are found to be used as weight of the digging stick by the present day Chenchus tribe of Andhra Pradesh[34]. So, we may infer, indirectly, from this evidence that this ring stone might had served the same purpose in Neolithic times also. In addition to that the ring stone might have been fitted in the spears also.

Another Neolithic type of tool, almond shaped celt, has been reported among the Neolithic implements of Tamluk, West Bengal[35]. This celt shows a convex cutting edge with nearly elliptical cross section. Iron blade of similar shape is found to be used by the present day Lhota, Sema and other Naga tribes[36]. The blade of the hoe simply tapers upwards from the edge. It is found on the front of the shorter limb of an angular handle, cut from the junction of two branches.

Another agricultural hoe called *akuwa* by the Sema Nagas[37] and 'ik' by the Adi tribe consists merely of a narrow strip of bamboo. It is made in the shape of a necktie. The central part of it is shaved down to form a sharp, flexible blade (Pl 32C). The ends are prolonged to cross and afford a hold for the hand. This lighter and handier implement enables the user to do more work within a given time and also preventing injuring the younger crops.

Another variety of Neolithic tool, which has been classed as 'axe with broad cutting edge' by A. H. Dani[38] is found to be used by the Kuki tribe of north Cachhar hill area. According to H. B. Medlicott[39] this type of stone implement was used as hoe until few years back. The stone was set into a wooden handle and used for tearing out weeds in the rainy season when the ground remains soft.

The Nishi or Dafla is one of the largest populated tribes of Arunachal Pradesh. This tribe uses a hoe called *lowdap* (Fig 15 B) for clearing weeds. The artifact itself provides in the form of notches in relation to its working edge. This *lowdap* might have its origin laid in the late Neolithic period[40], in farming operation. This assumption is attested by the presence of various such types of constricted stone artifacts which were definitely used with a haft.

Another agricultural implement, a chopper, is designated as *da* by the Nishis[41]. Its shape suggests that it might be a new adoption and functionally replaced the axe of the Stone Age. This Nishis also used a stone axe, *alengahey* (Plate 15C)[42] for cutting and felling trees. This type of stone axe commonly found world wide during Neolithic times.

Among the various kinds of celts, used by the Nagas, roughly triangular celt predominates[43]. Widest part of this tool lies at the edge. Another type of celt used commonly by the Nagas is Neolithic type of wedge shaped celt. This type of celt is bound to crooked sticks before use[44]. The Neolithic wedge shaped celt has a broad, straight, median and bifacially ground cutting edge. The butt end is flat and pointed. This end is chipped transversely and ground.

Survival of such rudimentary implement among different tribes is nothing but an indication of antiquity of these tribes. Contemporary culture of few such tribes is indeed archaic in nature, which has its anchorage in the past. It is found that all types of above described tools, which are carried down from Neolithic age are mostly related to agricultural operations. So, it will again be reasonable to conclude that eastern and north-eastern Indians would have been rudimentary farmers by Neolithic times.

In addition to the tools described above there are other tools whose functional aspect along with different hafting process could be conjectured from ethnographic parallels.

Chisel is a characteristic tool of Neolithic culture. This cylindrical tool possesses a sharp cutting edge with a flat or rounded butt end. This tool is used by present day tribes for splitting or smoothening plank of wood. So, similar service might have been availed in Neolithic times also.

In Neolithic period the use of hammer was pretty common at every zone. It was generally made on elongated pebble, with any working, hardly. In present context, an old Khampti tribe of Lohit district has been reported[45] as using similar fine hammerstones. Its handle is made of a creeper and lashed on.

Hutton described[46] a celt of white stone from Naga Hills. As it was nearly square in shape, it was not easy to reconstruct its hafting procedure. This confusion was also solved by another ethnographic parallel. Such celts might have been hafted like those of the Phom Nagas, who effectively haft their stone hammer by fixing it in a cane loop[47].

The confusion regarding hafting of lithic celts could be solved from other ethnographic analogy also.

Some shouldered celts are reported to be hafted adze wise in typical Neolithic fashion by present Naga tribe of Ledo[48]. The stone celt locally called *'ta-chumi'* is hafted by bending a cane rod round the tang to grip the later.

Another method of hafting is found among Lhota, Sema and other Nagas. In this method Neolithic type almond shaped celts are hafted by bending on to the front of the shorter limb of an angular handle, cut from the junction of two branches[49].

A very common Neolithic hafting practice in use in different parts of the world was the insertion of the blade into the holes near the ends of a club like handle[50]. This type of hafting is witnessed now-a-days in the Naga Hills in which the blade possesses a well defined single tang[51].

A number of hafting procedures of Neolithic tools have been presumed by Sankalia[52].

Thus it may be said that blades were hafted in various ways in that primitive age, depending on the geomorphological nature of an area. The survival of such typical Neolithic cultural traits among the aborigines dimly illumines the unseen world of the primitives.

In addition to stone, prehistoric man might have worked at least in wood, from which very satisfactory, fire-hardened points, cutting edges could be made. Besides such relics, there are other evidences which are freezed in the day to day activities of these tribes, indicating the archaic nature of the same.

The present day Mao tribe of Manipur prepare their clothes from a bark fibre[53], with the help of spindle whorl of pottery and edge ground stone perforator. Discovered terracotta spindle whorl from Chirand, Maner (Pl 28 A) might have been used for the same purpose in Neolithic period.

Ethnographic similarity also helps in providing a clue about the religious belief, active among the prehistoric people.

The Nagas worship plenty of stones, predominant among them are Neolithic type triangular and wedge shaped celts. Thus, it could be said that these Neolithic implements were not only used in economic pursuits but they had religious bearings too. Again miniature Neolithic celts of softer rocks also might have been worshipped. These smaller tools might have been used as pendants also.

Besides the Nagas most of the present day tribe worship natural objects like trees, hills, stones etc. From Chirand terracotta figurine of snakes, both hooded and coiled varieties, have been discovered. The Nagas, Khasis worship similar snake figurine, which is very popular is eastern India. Besides serpent worship, terracotta bull, amulet also implies their use in religious purposes.

It is evident that in countries like India where tradition persists, many ancient tool types may continue in unadulterated form being carriers of customs.

In making fire, the Santals often employ friction of wood method while the Savaras make it by drill[54]. Such primitive way of life definitely indicates a borrowing and carrying forward from the very early stage. It may be conjectured that Neolithic folk led almost similar livelihood, where natural, perishable material would have been used in the manufacture of daily usable objects. In the moist climate of eastern India these could have decayed.

This third World country is changing and developing rapidly and the tribals are also allured towards the amenities of modern civilized life and giving up their traditions. Thus there is urgent need of extensive study and documentation of such ethnographic information before its irretrievable extinction in near future.

References:

1. Kosambi, D.D. 1967 'Living Prehistory in India' *The American Review,* P.29.
2. Kramer, C.(ed) 1979 *Ethnoarchaeology : Implications of Ethnography for Archaeology,* New York, P 277.
3. Stainslawaki, M.B. 1974 'The Relationship of Ethnoarchaeology, Traditional and Systems Archaeology' in Donnan Clewlow(ed) *Ethnoarchaeology.* Losangels, P. 48.
4. Chang, K.C. 1967 'Major Aspects of the Inter Relationship of Archaeology and Ethnology' *CA,* Vol VIII, P. 227.
5. Bindord, L.R. 1968 Methodological Consideration of the Archaeological Use of Ethnographic Data in B.L. Richard and D.V. Irven (eds) *Man the Hunter,* Chicago, P.269.
6. Marshall, John 1931 *Mohenjodaro and the Indus Civilization*, vols. I, II, III London.
7. Mackay, E. 1938 *Further Excavation at Mohenjodaro and Chanhudaro,* Delhi.
8. Sengupta, G. 1991 Introduction in J.P. Singh and G. Sengupta (eds). *Archaeology of North Eastern India*, Delhi, P.XIX.
9. Haimendorf, F.1948 'Notes on the Stone Age in India', *MI,* XXVII (IV), P. 197.
10. Rao, S.N. 1973 'Excavation at Sarutaru: A Neolithic Site at Assam', *ME,* 1, pp 1-9.

11. Manoj Kumar 2003 An Ethnoarchaeological Study of Neolithic Culture of Northern and Eastern India (un-published Ph D thesis).
12. Goswami,M.C, S.K. Roy 1975 Note on Tangsa Hoe, Arunachal Pradesh, *BDA* vol IV, pp. 71-73.
13. *ibid.*
14. Manoj Kumar 2003 *opcit.*
15. Playfair, A 1909 *The Garos,* London, P. 39.
16. Lyall, Charles, 1908 *The Mikir,* London, P. 53.
17. Sangma, M.S.1991 *Op. Cit,* P. 304
    Rout, S.P. 1969-70 'Hand Book on Juang' *Adivasi* , vol XI, p. 6.
18. Ashraf, A.A 1990 *Prehistoric Arunachal: A Study of Prehistory and Ethnoarchaeology of Kamala Valley.* Directorate of Research Govt. of Arunachal, P. 110.
19. Sharma, H.C, A. A. Ashraf, H.C. Mahanta 'Ceramic Tradition in Brahmaputra Valley Assam: A Case Study From Ambari': Gauhati, pp. 87-96.
20. Hutton,J.H. 1968 *The Sema Nagas,* Oxford University Press. P. 66.
21. Chakrabarty, M, D. Mukherjee 1971 *Indian Tribes, Calcutta,* P. 48.
22. *ibid.*
23. Sinha,B.P. 1998 Continuity and Change from Neolithic to Chalcolithic Phase : A Case Study of Senuwar in C.P. Sinha (ed). *Facets of Indian Culture,* Patna, pp. 5-12.
24. Gurdon, Colonel 1907 *The Khasis,* London, P. 18.
25. *ibid.*
26. Balfour, H 1917 'Some Types of Native Hoes, Naga Hills' *Man,* July XVII, pp. 105-107.
27. Mill J.P. and J.H. Hutton 1929 Ancient Monoliths of North Cachhar, *JPASB,* XXV (N.S), P 295.
28. Roy Chowdhury, P.C. 1965 Bihar District Gazetteer, Santal Pargana, Patna, P. 47.
29. Hutton, J.H. 1931 *Census of India*, vol. I Pt. III, pp. 50-60.
30. *Ibid* pp. 50-60.
31. Sangma, M.S. 1991 Crafts and Technology of the Garos in J.P. Singh and G. Sengupta (eds) *Archaeology of North-Eastern India.* P. 304.
32. Medhi, D.K. 1983 'The Garos and their Material Culture : A Study Based On Ethnoarchaeological Approach, *ME,* VIII, pp. 70-76.
33. Mohanty, P. 2000 Mesolithic Culture and Ethnography of Keonjhar District, Orissa in P. Mohanty and K.K. Basa (eds) *Archaeology of Orissa.* P. 139.
34. Dasgupta, P.C. 1967 *Pragaitihasik Susunia* (Bengali), Directorate of Archaeology, West Bengal, P. 49.
35. Mukherjee, T. 1958 'Neolithic Implement from Tamralipta' *Indian Folklore,* Vol. I (III), pp. 42-44.
36. Balfour, H. 1917 *Op.cit.* pp 105-107.
37. Hutton, J.H. 1968 *Op.cit,* P. 66.
38. Dani, A.H. 1960 *Prehistory and Protohistory of Eastern India,* Calcutta, P. 53.

39. Godwin Austen, H.H. 1875 'A Celt Found At Khasi Hills of Shillong', *MSB*, P. 158.
40. Ashraf,A.A. 1994 'Anthropo-Archaeological Background of the Nishi Material Culture' *BDA*, VIII, pp. 105-113.
41. *ibid* P. 110.
42. *ibid* P. 110.
43. Hutton,J.H. 1969 *The Angami Nagas*, Oxford University Press. AppendixVII, P.403.
44. *ibid* P. 404.
45. *ibid* P. 404.
46. Hutton,J.H. 1924 Two Celts from the Naga Hills', *Man*, pp. 20-22.
47. *ibid* P.22.
48. Hutton, J. H. 1924 *Op.cit*
49. Balfour,H. 1917 *Op.cit*, P. 114
50. *ibid* P. 105.
51. Sankalia, H.D 1964 *Stone Age Tools. Their Techniques Names and Probable Functions*, Poona, pp. 102-103.
52. Narain, L.A. 1981 The Chalcolithic Settlement In Bihar in M.S. Nagraja Rao (ed) *Madhu, Delhi* P.54.
53. Singh, O.K. 1988 Aspects of Archaeology in Manipur in N. Sanjaoba(ed). M*anipur, Past and present*, Delhi, P. 103.
54. Roy, H.M 1986 Savara the Snake charmer, Calcutta, P. 27.

Conclusion

In the foregoing five chapters, we have tried to give a picture of the pattern of culture that might have prevailed during the Neolithic period in eastern and north eastern part of India. It is perspicuous that this period is characterized by a total development of many industries that have its root in preceding Mesolithic phase even. Although much controversy exists as to whether it should be appropriate to call this phase as a revolutionary phase or not but the fact is that the Neolithic culture is a watershed in human history, whose arrival marks one of the most profound changes in the pattern of his livelihood. At present Neolithic archaeology is enriched with sufficient Neolithic material evidences. An overview of Neolithic cultural phase of eastern and north eastern India reveals that West Bengal Neolithic in stratified context are yet to be found. Surface finds comes broadly from three clusters *i.e.* Subarnarekha complex, Kangsavati Complex and Susunia hill complex[1]. Most frequently found lithic assemblage of the Subarnarekha river valley are shouldered celt, bar celt, chisel, mace head and splayed axe while it is ring stone, adze, splayed axe in the Kangsavati valley and bar celt and ring stones at Susunia hill. Another region of concentration of technically more advanced Neolithic folk of West Bengal lies in the northern most part of the state *i.e.* Darjeeling district. Characterictic yield of this cluster are perforated celt and ornament.

In Jharkhand, the Chotanagpur plateau, Singhbhum, Santal Pargana and Ranchi districts are very rich in yield of Neolithic evidences which shows differential phases of development. The wholly chipped to wholly ground specimen of Borda-Bongara group represent an earlier phase over Dungi-Barudi Neolithic industry, which contains rare specimen of chipped to ground tools but some new tool types *viz* quadrangular axe and adze were added. This Neolithic industry of Singbhum bears affinity with southern peninsular and eastern Indian Neolithic character while Santal Pargana neoliths are an admixture of both Indian and South East Asian forms. Beyond this area, eastern Indian Neolithic community also flourished in the Gangetic valley area of Bihar at Chirand, Chechar Kutubpur, Senuwar, Maner and Taradih. These industries were predominated by celt, terracotta object, bone and antler tools, hand and wheel turned pottery which is quite developed than Chhotanagpur Neolithic industry. Among the five excavated sites, a comparatively earlier phase is represented by Neolithic assemblage of Taradih, Senuwar and Maner while Chirand and Chechar Neolithic culture is more advanced in typology as well as technology. In Orissa, districts of Keonjhar, Mayurbhanj, Sundargarh, Dhenkanal and Puri are mentionable in Neolithic context. Fourteen different types of Neolithic tools are classified on the basis of their morphology and probable function, among them axe, adze, chisel, adze, shouldered celt, hammerstone are mentionable besides pottery.

In Assam, north Cachhar and Kamrup districts are most important in yield of neoliths. Predominant yields of these districts are quadrangular axe, adze, shouldered celt and cord marked pottery. Neolithic industry of Garo hill district, Meghalaya are dominated by rounded butt flat celt, shouldered celt in association of crude, hand made, plain pottery. In Arunachal Pradesh occurrence of Palaeolithic tools in Neolithic context suggests a continuation of earlier tradition into Neolithic age. In Neolithic level an aceramic and a ceramic phase have been identified. Neolithic tool types include shouldered celt, double edged waisted axe, cylindrical hammer etc. The Neolithic tools of Nagaland were mainly shaped by pecking initially and finished by edge grinding technique. Main Neolithic tool compositions are represented by feebly shouldered celt and pointed butt celt.

Neolithic culture of Tripura is dominated by polished axe. Lithic assemblage of the state indicates a woodland ecology in the area. Manipur Neolithic industry is akin to South-East Asiatic axe-adze culture. Besides axe and adze, a variety of potsherds (corded ware, geometric ware and tripod ware) enhanced archaeological importance of the state. According to K.C. Chang[2] the corded ware culture might be associated with shifting cultivation. Thus an early appearance of Neolithic culture in Manipur would reasonably be cŏnjectured. Association of net sinkers, micro-blade and point with tripod ware culture suggests a hunting fishing and food producing economy during Neolithic culture of Manipur.

Absolute dates of Neolithic cultures of India (Except Southern India)

Site	Material	Culture	Date
Northern India			
Burzoham	Charcoal	Neolithic Period IA	2586 B.C.E.
Burzoham	Charcoal	Neolithic Period IB	2465 B.C.E.
Gufkral	Wood	Neolithic Period IA	1313 B.C.E.
Gufkral	Charcoal	Neolithic Period IB	2881 B.C.E.
Gufkral	Wood	Neolithic Period 1C	1860 B.C.E.
Belan and Vindhyan Regin (earliest dates)			
Koldihwa	Charcoal		6570, 5440 B.C.E etc.
Kunjhun	Charcoal		3530 B.C.E
Mahagara	Charcoal		1628 B.C.E
Lahuradewa	cultivated rice	Petriod 1A	6442 B.C.E
Lahuradewa	Wood Charcoal	Period 1A	6376 B.C.E
Tokwa	unknown		6591 B.C.E
Tokwa	unknown		5976 B.C.E
Tokwa	unknown		4797 B.C.E
Jhusi	unknown		7477 B.C.E
Jhusi	unknown		6196 B.C.E
Eastern India			
Chirand	Charcoal	Neolithic	3338, 3213, 1740B.CE. etc. (earliest dates)
Senuwar	unknown	Neolithic – Chalcolithic	1770$\pm$120, 1660$\pm$120 B.C.E. etc.

Site	Material	Culture	Date
North Eastern India			
Kachuani (Dibrugarh)	Charcoal	Neolithic	2210±140 B.C.E
Nangpok Keithelmanbi (Manipur)		Neolithic	4460±120 B.C.E
Khas Kalanpur (Tripura)		Neolithic	3350±150 B.C.E

Relative dates of Neolithic cultures of Eastern and North Eastern India		
Barudih	–	1100-700 B.C.E
Golbai Sasan	Neolithic	1600 B.C.E.
Kuchai	Neolithic	1000 B.C.E.
Napachik	Neolithic	1450 B.C.E.

It is evident that at present there is no dearth of Neolithic material resources in this part of India and if these entire collections could be interpreted, it could reveal a clearer scenario of Neolithic culture of eastern and north-eastern India. Unfortunately most of these archaeological materials are found from surface and these unstratified tools are hardly much helpful for this purpose. Unfortunately the number of systematically excavated sites over the whole region is very insignificant, which is the greatest enigma in deducing the proper stratigraphical or contextual sequence of the Neolithic materials of eastern and north-eastern India. So, assigning date to Indian Neolithic period is highly tenuous. Besides that very few Indian Neolithic cultures could be dated in absolute methods. Orissa is still lacking an absolute date for its Neolithic period.

These meagre samples resulted in somewhat approximate fixation of age of the culture. Due to lack of good number of absolute dates from carefully controlled specimens, we are compelled to look for substitute relative dates.

Vishnu Mittre[3] has suggested a time bracket of 4,000 - 3,000 B.C.E. for Chirand Neolithic level. The period IA of Senuwar is pushed back to the beginning of 3rd millennium B.C.E. and period IB to 1500 B.C.E in relative dating.

It is almost an established fact that north-eastern Indian Neolithic was much indebted to East and South-East Asian Neolithic. Due to paucity of direct dates, eastern and north-eastern Indian Neolithic cultural specimens are dated in terms of dated Neolithic object of East and South East Asia.

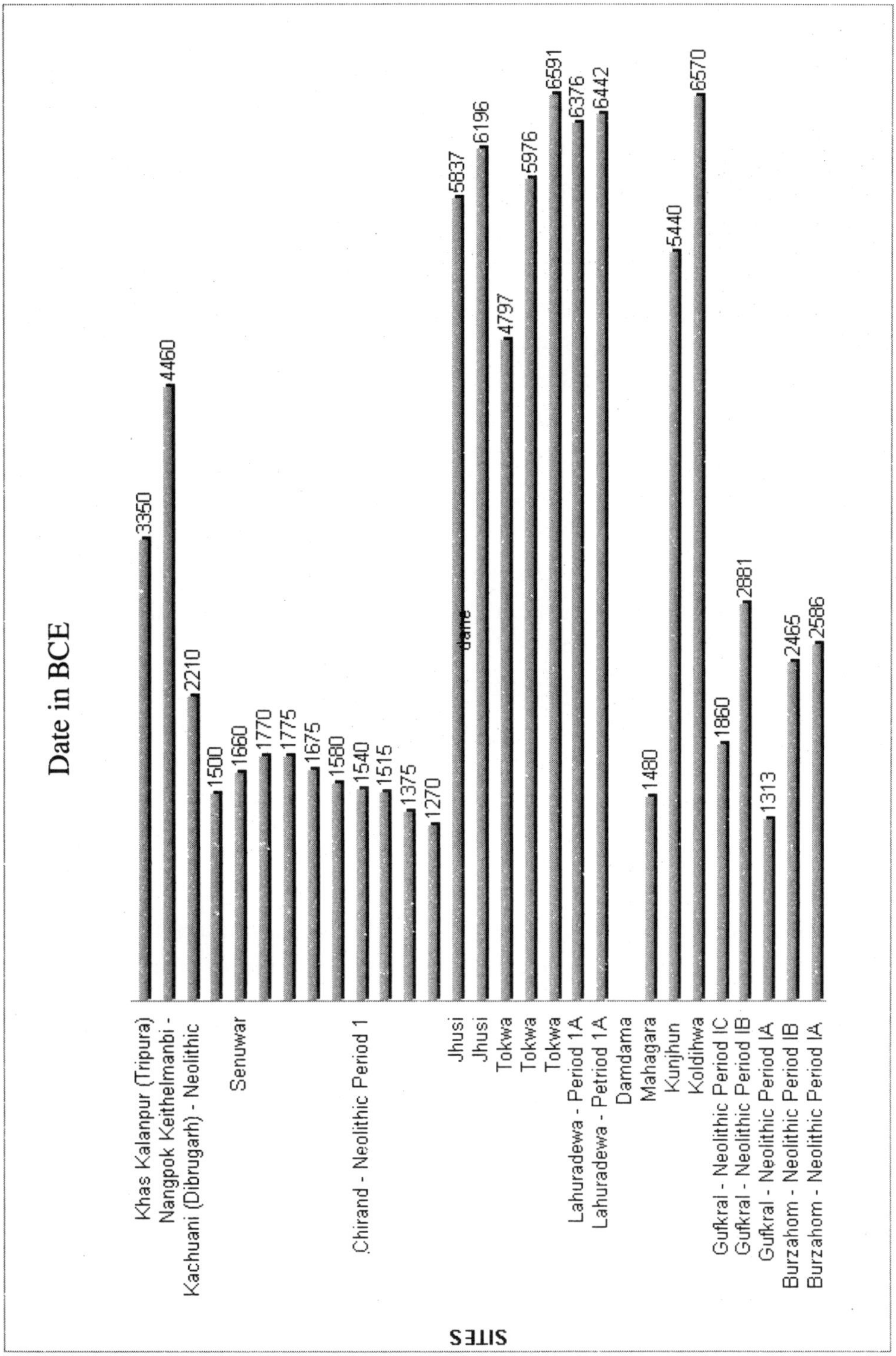

Fig.16 : Date Chart

Shouldered tool is a characterisitc tool of north East Indian states. It occurs in eastern India also. This tool type is simultaneoulsy reported from East Asia as well. Splayed edge axe of Assam is another tool found from Burma, Yunan and Malaysia. Again on typological ground Manipur Neolithic industry belongs to round and quadrangular axe-adze culture of South-East Asia. Rectangular axe of Assam is very similar to that of north-west China. The gouze adze found from Assam shows its resemblance with that of Somrongsen culture of Combodia. Driem is of opinion that north east Indian Neolithic is more akin to Chinese Neolithic than that of South East Asia. (Driem G V 1998 Neolithic Correlates of Ancient Tibeto Burman Migration, in Blench and Spriggs (eds) *Archaeology and Language Ii: Archaeological Data and Linguistic Hypotheses* pp. 67-102.) Typologically similar tools of these regions must have been used in close chronological horizon. Sankalia[4] opines that late Bacsonian culture of Vietnam is the source of Assam Neolithic culture and suggested a date of 5,000 B.C.E. for its early Neolithic phase while H. Geldern[5] suggested 2500- 1500 B.C.E. for South-East Asian Neolithic cultural period. Some others claimed that South-East Asian Neolithic period continued upto the appearance of Chinese glazed ware while some authorities prefer to stretch the time upto the appearence of iron. According to E.C. Worman[6]. East Asiatic potteries and early types of stone tools should be placed before 3,000 B.C.E. and if a time is added to this date to reach to India, a date not earlier than around 2,000 B.C.E. could be postulated. K.D. Banerjee opined that eastern Indian Neolithic desrves it origin from Hoabinhian culture[7].

Neolithic date is also estimated on the basis of resemblance between pottery forms. The cord-marked pottery of north-eastern India is comparable to the Neolithic pottery of Jomon culture of Japan, Yangshao and Langshanoid culture of China and Hoabinhian culture of Vietnam, thus pushing back the antiquity of Indian counterpart to early Holocene period. Three phases of Neolithic *i.e.* corded ware culture, tripod ware culture and geometric ware culture of Manipur were also recognised in China. The geometric ware phase is dated to 1500-700 B.C.E. in Chinese context and Tripod vase from Ban kao cemetry Thailand is dated around 1,800-1,400 B.C.E.

Not only in north-eastern India, South-East Asiatic influence could be traced as far as Orissan territory in Neolithic period. Shouldered axe tradition diffused along the coast of Orissa upto north eastern Andhra Pradesh while triangular axe diffused through Bihar upto Uttar Pradesh. H. Geldern[8] argued that shouldered adze reached Orissa from South-East Asia through land route via north-east India during 2500 B.C.E. -1500 B.C.E. Besides that Neolithic material assemblages of Chirand, Senuwar and Kuchai suggest a definite link with Neolithic Thailand. Shouldered tool, chisel found from Sulabdihi, Orissa, is very much similar to the pick adzes of South-East Asia. Few Neolithic collection of jade from Tamluk, West Bengal, are similar to those found from Malay, Yunan, China - all of which are river based culture. Further the burnished pottery and bone tool found from Indian Neolithic culture shows resemblances with that of Hoabinhian culture.The artifact assemblage of Santal Pargana and north-east India also shows resemblances. The fossil wood industry of Tripura very well ties up with the late Anyathain-2 level of Myanmar and Lalmai of Bangladesh. At Daojali hading rounded butt axe of Indian type and shouldered tool of South-East Asiatic variety occurred in the same stratigraphic level.

Apart from material resemblances, the presence of Austro-Asiatic linguistic group in north-eastern Indian population also proves South-East Asian contact. The South- East Asian type Mongoloid features Mon-Khmer groups are still found among the Mundari speaking Orissan people of today.

So, it could be said that though Neolithic industry of north-eastern India is largely dominated by East and South-East Asiatic types, its impact was felt far beyond the territory. Simultaneously, it should be remembered that such trait infiltration was never a one way traffic and it could not be limited by any political border line. In northeast Indian Neolithic cultural repertoire, along with the East and South-East Asiatic variety, east Indian and south Indian cultural traits were also assimilated. Discovery of microliths and plain hand made pottery from Selbalgiri, Meghalaya shows influence of south Indian Neolithic culture over that of north-eastern India. Again south Indian Neolithic trait in eastern Neolithic could be demonstrated by the presence of pointed butt axe in many eastern Indian Neolithic sites. Co-existence of pointed butt axe, and eastern Asiatic neolithic type in Chotanagpur Plateace led F.R. Allichin to assume that this plateau formed a contact zone between Indian and East Asian Neolithic character[9].

We have traversed a long path after the time of the neolithians, mean while most of the relics of that remote past could not survive the ravages of time. To know about these early men and his culture the present ethnographic model has helped greatly to interprete the past archaic samples. It is found that many tribes of present day use various tools which retain their forms of Neolithic time. Though the raw material does not always exist in lithic form today and transformation from stone to other materials might have incorporated due to the aquintance with more effective and suitable one, without changing the functional purpose. Such similarity in material milieu is an outcome of the uninterrupted development of human culture in the region.

Now if we look into the overall Neolithic cultures of India it could be found that all the regional Neolithic bears some distinguishing features of their own while some features are also common with others. Neolithic people of both north and mid- eastern India used a variety of ground stone tools and in north-eastern India, shouldered celt and quadrangular adze most commonly occurred. In this period, bone was also utilized by peoples of some places *viz*. Chirand, Chechar, Gopalpur, Golbai Sasan etc. in eastern Neolithic zone, Mahagara in Vindhyan Neolithic zone, Piklihal and Utnur in southern Neolithic zone, Gufkral and Burzahom in northen Neolithic zone. Bone tools are conspicuously absent in any neolithc site of north eastern Indian sites. In a number of Indian Neolithic cultural assamblage besides bone tools, microliths are found either to precede (Sanganakallu, Chopani Mando etc.) or to be associated (Koldihwa, Mahagara) with Neolithic artifacts. In eastern India from Kuchai, Orissa and in north eastern India from Selbalgiri (Meghalaya), a microlithic phase underlying the Neolithic level have been discovered, but in northern Neolithic no evidence of microlith in relation to Neolithic culture could be traced. A trial digging in Riat-Turein, Jaintia Hill district, Meghalaya yielded a continuous cultural succession from Palacolithic to Neolithic, which deserve more investigation.

Cord impressed pottery is evident from Vindhyan and north eastern Indian Neolithic sites but the north eastern variety differ in colour and design from that of the former and represent a later stage of development. Again in the incised reddish sherd from Barudih, kankar is used as tampering material which bear affinity with coarse grit tampered red ware sherd of Kuchai.

In settlement history it could be said that by the Neolithic times, people of eastern and north-eastern India started raising their dwellings. Their huts were sometimes circular (Plate 30 A) or quadrangular in ground plan and were made up of either mud or wattle and daub with mud plaster. In addition to that from ethnographic evidence it could be conjectured that perishable, but easily available local materials like timber, bamboo, reed, leaves might have been used in raising temporary structure specially by the shifting cultivators of northeastern Indian hill people, like most of the present day hill tribes. Mud fortification wall encircling the habitation could be traced from the discoveries at Chirand. Another site named Sachima near Kohima, a stone fortification wall has been found[10]. At Mahagara in the Belan valley, settlement was constructed in a depression enclosed by natural embankment of cemented gravel.

In northen Neolithic culture, people did not built such structures, initially (period IA, IB), rather circular or oval and square or rectangular pits in the ground were used as dwelling places. This pit dwelling are not found in other eastern and North-eastern Indian Neolithic site of India except Chirand but at Neolithic sites of Levant, Iran, Afghanistan, Pakistan, China, Japan as well as Europe. In Chalcolithic Inamgaon as well as few Harappan sites have yielded this pit dwelling structure. In its matured phase (period 1C) of northern Neolithic culture mud brick structures were raised. In southern Neolithic, at Paiyampalli only such pit dwelling have been found, otherwise Neolithic huts of south India had mud plastered wall with lime, clay or mud silt and dung made floor. Wooden posts were used to support the concial thatched roofs of such huts. In Vindhyan Neolithic complex an open air sedentary settlement has been evident from proto Neolithic Chopani Mando. Such dwellings had stone paved floor with wattle daub screen. The huts at Mahagara, in the Belan valley, are well arranged and uniform and two to three such huts make up a house unit. Among them one hut at least was used for living purpose while others are used for manufacturing and cooking etc[11].

The settlements were mostly selected in highlands on the river bank or adjacent to forest as Neolithic folks were dependent largely on gathering of forest produces and riverine creatures along with hunting and farming. In north-eastern India all the Neolithic sites are located in hilly areas. Selection of highland was due to avoid flood. In Vidhyan Neolithic, main sites were found along the river valley, adjacent to or along the horse shoe lakes. In south India neolithians used to live on the tops, slopes and to the feet of castellated granite hill characterised by background like rock shelters or caves, plateaus, water creatures, wild fauna, raw material etc. All the Neolithic sites of northern Neolithic zone are located on the elevated flats of the karewas often overlooking streams and lakes.

So far as the subsistence pattern of Neolithic culture is concerned, it could be said that in eastern India, recovery of large number of animal bones from Chirand and Gopalpur are indication of both hunting and domestication of animals. Present use of many Neolithic implements *viz.* shouldered celt, ground celt, ring stone etc.

in agricultural purpose by various tribal groups of this region and discovery of carbonised rice, lentil, mung, pea etc. as well as ground stone tool, saddle quern etc. among the Neolithic assemblage of different parts of eastern and north eastern India attest some sort of agricultural practice. In south Indian Neolithic, subsistence economy ranged from collecting-gathering, hunting, fishing to animal husbandry and rudimentary agriculture.

In northern Neolithic culture at Gufkral people were specialized in food gathering, producing and also and stock raising to some extent, but at Burzahom direct evidence of cereal farming could not be established though occurrence of harvester and macehead indicate a knowledge of husbandry. Belan valley is the only nuclear region of India where continuous stages of transformation from food gathering to developed food producing economy has been identified.

Remains of bones and horns of cattle and goat from excavation at Chirand imply that the society domesticated these animals. In other eastern Indian Neolithic sites at Golbai, Gopalpur (Orissa) (Plate 18) and Laljal and Bhaluksonda cave (West Bengal), cattle might have been domesticated. South Indian Neolithic cultures might have good number of domestic animals. Animal remains of sheep, goat and cattle are the most widely occurring species of these sites. In Neolithic Gufkral goat and sheep and in Burzahom dog and goats were domesticated. Among the Neolithic sites of Vindhyan complex, Mahagara and Chopani-Mando proved the domestication of cattle, sheep, goat and horse. In north-eastern India no eivdence of domestication of plants and animals has yet been found out. Bagor in Rajasthan Loteshwar in Gujarat etc have produced evidence of local development from hunting gathering to domestication of both plant and animal[121] in microlithic level.

As it has been seen earlier (chapter III) at least four main types *viz.* red ware, grey ware, cord marked / grey coloured and burnished grey ware, were dominant pottery of eastern and north eastern India. In south Indian Neolithic the principal wares were grey and dull red ware. Black on red ware was known to the people of this area late in their career. In Vindhyan Neolithic at Chopani Mando, earliest recorded evidence of the manufacturing of pottery in India is recorded. Hand made pottery of this region are cord impressed and burnished ware. The Burzahom Neolithic of northern India is represented by hand made, mat impressed pottery. In succeding phase (period 1C) black burnished ware made its appearence. In West Bengal, in the northern hilly zone, the Neolithic culture is completely devoid of ceramic trait. Such an aceramic Neolithic level has also been found from Parsi Parlo in Arunachal Pradesh.

Discovery of bones of birds and animals imply the prevalence of hunting. The settlement near the forest also proves hunting as well as food gathering occupation of the neolithians. The large number of fish bones from Chirand and probable fishing net from Teesta river site from Neolithic context also suggest fishing as an economic pursuit of neolithc people. The discovery of terracotta wheel with perforation in the center (Chirand, Maner Plate 28 A) and the spindle whorl from Chirand and present use of similar edge ground stone perforator and spindle whorl for dress making from fibre like materials by the Mao tribe of Manipur (Plate 28 B) strongly suggest probability of knowledge of dress making in Neolithic culture. Neolithic potteries with mat, basket impression from Neolithic eastern and north- eastern Indian sites indicate that basket and mat making existed during the Neolithic time.

Difference of profession might have sowed the seed of social stratification by this period. Discovery of series of hearths from Chirand in Bihar indicate existence of some sort of community life in Neolithic period. In Belan valley Neolithic or proto Neolithic, a number of hut structures, closely situated to each other are also indications of community life. In Mahadaha cattle pen amidst several house units suggest cattle must have been community property. Community life could be infered from the practice of hunting, fishing as well as shifting cultivation also. It could be presumed that like the present day tribal groups mutual co-operation was the main feature of the Neolithic society also.

So far as the dietary habit of the Neolithic people of our country is concerned, it can be deduced from the discovery of charred animal bone, bones of fish, bird etc., that people of the period relished non-vegetarian dish, besides rice, wheat, moong, ragi, barley, pea etc. In south Indian and Vindhyan Neolithic culture almost similar dietary habit could be conjectured. From pollen evidences it can be said that the people used to consume both cereals and pulses. Sheep, goat and cattle were domesticated for their milk as well as meat. From Burzoham, fossil evidences of animals suggest that they were utilised for milk and meat.

Some sort of artistic activities by these early men in the form of rock painting, designing of ornaments, manufacturing of terracotta figurine etc. suggest some sort of art activity in the society. In eastern India, Orissan Neolithic phase has witnessed rock painting and engraving from a number of sites such as Yogimath in Kalahandi district, Vikramkhol and Ulapgarh, Sambalpur district, Manikmunda and Ushakothi at Sundargarh district[13] (Plate 20A). In Bihar, rock paintings have been found at many rock skelters. In these rock paintings animal figures, human figures and different symbols are represented in red ochre colour and white[14] (Plate 20B). Few engravings are also evident from this site. In south Indian sites red ochreous rock paintings of tortoise, deer, bull etc. have been found. In northern Neolithic two engraved stone slabs are mentionable in this context.

Recovery of terracotta bull, female and snake figurines, amulet etc. in archaeological context suggests that the Neolithic people observed some sort of religious activities. Nature worship among present day tribals indicates to a belief in animatism as well as image worship which might be a survival of Neolithic practice. A ritual origin also seems plausible in case of miniature celts made out of softer materials.

In disposal of dead, cremation might have been practiced. Though concrete evidence is yet lacking. At Mahadaha, in the Ganga valley a cemetery has been found within the habitation complex. The graves are oriented in east-west as well as west-east direction. From some graves, burial goods have also been obtained. In archaelogical context Haimendorf associated megalithic burial practice of north-eastern India with Neolithic culture though megaliths are in general correlated with iron using community. In southern Neolithic, comparatively good number of fossil evidence have been reported, which tells that both pit and urn burial were practiced. In Kashmir valley no Neolithic burial were found in period IA, though both inhumation and secondary burial were practiced in period IC. The most notable feature of this Neolithic culture is the presence of animal burial practice.

Besides these, few other features could be inferred from some discoveries made in eastern and north-eastern India *viz*. existence of some sort of trade and transport could be postulated from the discovery of the toy cart wheel from Chirand. From Assam, Nagaland, etc. stone tools of jadeite rock have been found. As jadeite is not locally available in this region, it must have reached this area from foreign sources, through some sort of trade. The ocurrencc of shouldered celt, Jadeite made tools, burnished pottery and bone tools of Tamluk and shouldered tools of Orissa might have reached there as a result of maritime ventures from south China and Formosa. Favourable wind blows from south-China, Formosa towards Bengal-Orissa coast and its exploitation in historical period is well known.

From this above discussion, it could be said that in India, Belan and Ganga valley Neolithic complex might be one of the earliest of its kind where rice and at least cattle were first indigenously domesticated. In Gangetic Bihar Neolithic might have developed in association with different Neolithic cultures of India. Another centre was the mineral rich Chhotanagpur plateau where habitation could be traced right from lower Palaeolithic period. South-western plateau fringe area of West Bengal and northern part of Orissa might be extended home of this group of folks. The south coast of Orissa and West Bengal might have experienced some South-East inspiration along with the north-eastern India.

Except the river-valley and coastal settlement the contribution of Neolithic in the succeeding cultures are not evident. In the river spread of such valleys, farming villages with knowledge of copper technologies promoted to Chalcolithic Phase. Similar cereals were cultivated. The next period saw the introduction of iron and resultant surplus leading to the emergence of ranking and wealth competition as postulated by Hayden.

In Neolithic cultural period, in eastern and north-eastern India, people might have settled in river valleys and coastal areas. In this areas thick occupational deposit proves a larger duration of Neolithic settlement. In this sedentary life, reduced group mobility as well as assured food supply to some extent might have allowed shorter child spacing causing demographic disequilibrium as conceived by Binord. Few studies of palaeo-environment of north-eastern and eastern India could not suggest any greater fluctuation of climate which necessitates the innovation like agriculture as proposed by Childe. From the few investigation of the early Holocene climate of the present study area, it could be suggested that eastern India faced a warm humid climate during this time. In north-eastern India also palyno-stratigraphical studies of the Tertiary sedentary formation suggest prevailing of humid and tropical weather there. In the wild rice growing Koraput-Jeypore tract of Orissa, presence of cultivation indicator like Croton, theaceae from Brahmaputra basin and discovery of domesticated crop from Neolithic level of Bihar and Orissa amply corroborate Binford's theory of origin of agriculture in marginal areas of wild crop growing centres. Flannery's postulate of pre-adaptation of agricultural criteria like ground rubbed stone, crude ceramics are evident in preceding phase of Neolithic (*i.e.* mesoithic) culture in India. The grinding technology might at first used mainly (but not necessarily solely) for milling ochre. So, the technology was known and when man eventually turned to cereal grasses, he had only to adopt and expand a pre-existing technology in order to deal with grain processing[15]. Recent studies show that all the traits regarded as diagnostic to Neolithic culture are directly evidenced in Mesolithic level of India itself (settlement evidence from Mesolithic Sarai Nahar

Rai, Mahadaha, and Damdama, pottery from Mesolithic Mahadaha, domesticated rice from Mesolithic level of Bajpur, Orissa, domesticated animals from Mesolithic Ganeshwar, Bagor, Tilwara, Langhnaj etc.). It can be said that in Neolithic phase these pre-invented technologies were flourished only, more investigation can only establish these facts.

Thus, in this study we have seen that sufficient material remains are discovered, but inconsistencies of occurence and proper analysis demand more archaeological activity in this region. Archaeologically potential areas need more attention from authorities. More unfortunate are those cases which are not properly reported and published. Again in some cases it is not possible to reach to the original specimen, either because of preservational shortcomings or loss during transportation. A specimen which overcomes these entire hurdles, further suffers from wrong identification or partial description.

Besides these general contrivances, eastern and north-eastern Indian archaeology is suffering from different ailments.

Inspite of discovery of sufficient Neolithic evidences from south western plateau fringe and northern mountainous region of West Bengal, not a single Neolithic site has yet been excavated. More concerted excavation is likely to bring Neolithic occupational strata in this region.

Compared to other areas, archaeologically Bihar is relatively better documented, still further research should be undertaken to realise the preceding phase of Neolithic level of Chirand and Chechar. Again, at Barudih the sample of carbonised rice has created some dispute, because of its association with iron pieces, it is hardly justifiable to relate it with Neolithic assemblnge. Rich in material resources, Bihar and Jharkhand have long prehistoric sequences. Still it is quite surprising that from the highly potential Chhotanagpur belt, no authentic Neolithic site has been excavated till date. Inspite of a very well presented Neolithic horizon, Mesolithic phase in Bihar is yet to be exposed. The most outstanding Neolithic site of Eastern India, as well as whole of this subcontinent, Chirand has yielded a very rudimentary Chalcolithic assemblage following a flourishing Neolithic stage which calls for an investigation.

Discovery of large number of cultural fossils from Orissa, during last few years, has considerably changed the prehistoric scenario of the state. Non availability of ^{14}C dates of the excavated materials is the biggest lacunae of Orissan Neolithic. The importance of the state as one of the centres of origin of agriculture is also not considered seriously. At Kuchai excavator could not touch the natural soil due to its high water table. Thus this site and its adjoining area need re-investigation. Palaeo-environmental studies are also pending in this region.

The characteristic hand made cord impressed Neolithic pottery of north-eastern India is dated to early Holocene on the basis of its similarity with Yangshao, Langshanoid and Hoabinhian cultures but excavation of Sarutaru and Marakdola in Assam suggest that Neolithic age of the region may not go even beyond the Christian era. Again the Neolithic culture of Kanaigaon reserve forest in Dibrugarh has revealed a date in the circa 6th century C.E. So, is it feasible to continue such a prolonged prehistoric phase?

Unanimity in opinion seems to reign over Meghalaya archaeology, for years. In Manipur archaeology, more initiatives to shoulder O.K. Singh are desired for the development of the same. Nagaland, which was in the limelight of archaeological research during the British period fails to carry forward the tradition after wards. Mizoram is yet to prove its archaeological potential. Although Arunachal Pradesh stands as a formidable settlement zone of Neolithic period, she remained unhighlighted due to lack of initiative. Prehistorically, Tripura virtually remained a *terra incognita* hitherto, except some recent efforts. Problem oriented excavation can bring forth valuable information and save unnecessary wastage of fund.

These spectacular, unearthed remains which have been described and classified extensively, call for more analysis, now. More and more work in this untrodden field through multi-disciplinary approach will solve the puzzle of human history and enrich Indian archaeology.

References

1. Chattopadhyaya, Sengupta *etal* 2005 *An Annotated Archaeological Atlas of West Bengal*, Vol I, Delhi, P.72
2. Chang, K.C 1974 'Ancient Farmers in the Asain Tropics, Major Problems for Archeological and Palaeo Environmental Investigation of South East Asia at the Earliest Neolithic Level, In A.K. Ghosh (ed) *Perspective In Palaeo Anthropology,* Calcutta, pp. 36-43.
3. Mittre, V.1972 'Neolithic Plant Economy at Chirand' *PB*, XIX (Iii), PP 18-22.
4. Sankalia, H.D 1981 'From History to Prehistory' In V.S. Srivastava (ed) *Cultural Contours of India,* Delhi. pp. 59-71.
5. Geldern, H. 1940 Prehistoric Research in the Netherlands Indies in P. Honig (ed) *Science and Scientist*. P. 138

6. Worman, E.C. 1949 'The Neolithic Problem in the Prehistory of India', *JWAS,* XXXIX (B), pp. 181-201.
7. Banerjee,K. D. 1987 Stone Age Migrations in India in D.P. Chattopadhyay (ed) *Archaeology and History* N.Delhi, pp. 41-48.
8. Geldern, H. 1940 *Op.cit,* P. 138.
9. Allichin, F.R. 1962 The Neolithic Stone Industry of Santal Parganas *BSOAS,* XXV (ii) pp. 306-330
10. Manoj Kumar 2003 An Ethnoarchaeological Study of Neolithic Culture of Northern and Eastern India (Unpublished Ph D Thesis)
11. Sharma. G.R 1980 *Beginning of* Agriculture *Epi-Palaeolithic to Neolithic: Excavation at Chopani-Mando, Mahadaha and Mahagara,* Allahabad.
12. Shinde V.S. *et al* 2004 'Emergence, Development and Spread of Agricultural Communities in South West Asia', in Yasuda (ed) *Origin of Pottery and Agriculture*
13. Pradhan S. 2001 *Rock Art in Orissa.*
14. Neumayer, E. 1994-95 Rock Painting from Hazaribagh *Puratattva* 25, pp. 80-84
15. Pal, J.N. 2003 'Mesolithic Settlement in the Ganga Plain', *ME,* xix. P.94

Bibliography

Agarwal, D.P. 1982, *The Archaeology of India,* Scandanavian Institute of Asain Studies, Monograph Series no. 46, London.

Agarwal, D.P, J.S. Kharkwal 2002 *South Asian Prehistory* : A Multi-Disciplinary Study, Delhi.

Ahmed, E. 1965, *Bihar : A Physical, Economic and Regional Geography,* Ranchi P. 14

Allchin, B, 1966, *The Stone Tipped Arrow: Late Stone Age Hunters of the Tropical Old World,* London.

— 1995, *Living Traditions, Studies in Ethnoarchaeology of South Asia.* Oxford and IBH Publishing Co.

Allchin, B, and F.R. Allchin, 1968, *The Birth of Indian Civilization,* London.

— 1982, *Rise of Civilization in India and Pakistan,* Delhi.

Allchin, F. R, 1960, *Piklihal Excavation,* Hyderabad.

— 1961, *Utnur Excavation,* Hyderabad.

— 1962, 'The Neolithic Stone Industry of Santal Pargana, *BSOAS,* XXV(11), pp. 306-330.

— 1963, *Neolithic Cattle Keeper of South India : A Study of the Deccan Ash Mounds,* Cambridge.

Allchin, F. R and Dilip K. Chakrabarti (eds), 1979-1997, *A Source Book of Indian Archaeology. I, II,* N.Delhi

Anderson, C.W. 1917, 'Prehistoric Stone Implements in Singhbhum District', *JBORS,* III, pp. 349-362.

Ashraf, A. A. 1990, *Prehistoric Arunachal, A Study on Prehistory and Ethnoarchaeology of Kamla Valley,* Directorate of Research, Itanagar.

— 1994, Anthropo-Archaeological Background of Nishi Material Culture, *BDA,* VIII, pp.105-113.

Assam Research Society, 1984, *A Source Book of Archaeology of Assam and Other States of North eastern India.* Gauhati.

Avebury, Lord, 1900, *Prehistoric Times as Illustrated by Ancient Remains,* London.

Balfour, H. 1917, 'Some Types of Native Hoes, Naga Hills' *Man,* July XVII, pp. 105-107.

Ball, V. 1865, 'On Stone Implements Found in Bengal' *PASB,* pp. 127-128.

— 1867, 'Note On Stone Implements Found In Bengal', *PASB,* P. 143.

— 1870, Stone Implements Discovered in Singhbhum', *PASB,* P. 268.

— 1874, 'On an Ancient Perforated Stone Found in Satpura Hills', *PASB,* pp. 96-97.

— 1875, 'On Some Stone Implements of the Burmese Type, Found in Pargana Dhalbhum, District of Singhbhum, Chhotanagpur Division', *PASB,* pp.118-120.

— 1876, 'On Stone Implements Found in the Tributary States of Orissa', *PASB* pp. 122-123.

— 1878, 'Two Stone Implements from Parisnath Hill (District Hazaribagh)', *PASB,* P. 125.

Banerjee, R.D., 1924-25, 'Neolithic Implements from the Abor Country', *An.Rep.ASI,* P. 102.

Banerjee, B, 1972, 'Morphological Regions of West Bengal', *G.R.I,* XXVI(Iv),

Bareh, H, 1967, *The History and Culture of Khasi People,* Shillong.

— 1970, Nagaland District Gazetteer, P.6.

Barren, Lt. 1872, 'Note on Stone Implements from the Naga Hills', *JRAI,* pp. 61-62.

Basa, K. K. 1994, *Problems and Perspectives in Archaeology of Orissa*, Bhubaneswar.

Basa, K. K., P.K. Mohanty, 2000, *Archaeology of Orissa,* Vol. I, II, Delhi.

Beeching Capt. 1868, 'Note on Some Stone Implements Found in the District of Singbhum *PASB,* P. 117.

Behera, P. K. 1992, 'Pebble Tool Component of Bonaigarh Neolithic Complex, Orissa', *ME,* XVII, pp.57-63.

Burkitt, M.C. 1925 *Prehistory*, Cambridge

___ 1926 *Our Early Ancestor : An Introductory Study of Mesolithic and Copper Age Culture in Europe and Adjacent Regions*, London. 1949

___ 1949 *The Old Stone Age,* Cambridge.

Bhattacharya, D.K. and S. L. Malik, 1986, *Aspects of Human Ecology,* Delhi.

Bellwood, P. 2005 *First Farmers : The Origin of Agricultural Societies,* UK.

Bhattacharya, D. K. 1992, 'Is Prehistory Dead in India?', Paper Presented in Panchanan Mitra Memorial Lecture.

Bhattacharya, D.K. 1998, *Outline of Indian Prehistory,* Delhi.

Bhattacharya, T.K. 1979, *The Adis-Origin and Migration,* Calcutta.

Bhattacharya, K, Chanda, S. 1987-88, 'Late Quaternary Vegetational History, Palaeoecology and Bio Stratigraphy of Some Deposit of Brahmaputra Basin, Upper Assam, India, *JP,* XIII, pp.225-237.

Bhattacharya, M. 1982, 'A Note on Recent Explorations in the Upland Bengal in R. K. Sharma (ed) *Indian Archaeology, New Perspective,* Delhi, pp 48-52.

__ 1983, 'On The Recent Findings of Ground and Polished Stone Industries of South West Bengal' In B. N Mukherjee *et al,* (ed) *(Sri) Dinesh Chandrika, Studies in Indology,* Delhi, pp. 103-107.

Bhowmick, K. L. 1971, *Tribes of India, A Profile of Indian Ethnology,* Calcutta.

Bhowmick, P. K. 1963, *Lodhas of West Bengal: A Socio-Economic Study,* Calcutta.

Binford, L. R. 1962, 'Archaeology as Anthoropology', *AA,* XXVIII (II), pp. 217-225.

__ 1965, 'Archaeological Systematics and Study of Cultural Process', *AA,* XXXI, pp. 203-210.

__ 1968, Methodological Consideration of the Archaeological Use of Ethnographic Data In B.L. Richard and D.V.Irven (eds) *Man the Hunter,* P.269.

Binford S.R and L.R Binford, 1968, *New Perspectives in Archaeology,* Chicago.

Biswas, M.K. 1980, Animal Remains of Chirand, Bihar Records of *Z.S.I.* LXXVI.

Bodding, P.O. 1901, 'Ancient Stone Implement in the Santal Parganas' *JASB,* pp. 17-22.

__ 1904, 'Shoulder-Headed and Other Forms of Stone Implements in the Santal Parganas', *JASB,* pp. 27-31.

Bopardikar, B.P. 1972, 'Prehistoric Investigaticn, Dapabhum, Scientific Survey Expedition (NEFA) *Arch. Cong. and Seminar Paper,* Nagpur University, pp. 1-8.

Bose, N.K. 1928, 'Age of Boulder Conglomerate at Kuliana, Mayurbhanj', *P.I.S.C.* (28th Session), P. 212

___ 1948, *Excavation in Mayurbhanj,* Calcutta University.

Braidwood, R.J. 1940, 'Prehistoric Research In Mayurbhanj', *SC,* VI, pp. 75-79.

___ 1958, 'Near-Eastern Prehistory', *Science,* 127, pp 1419-1430.

___ 1967, *Prehistoric Man,* Chicago.

Braidwood, R. J and B. Howe 1960, *Prehistoric Investigation in Iraqui Kurdistan,* Chicago.

Brahmachari, U.N, and S.C. Brahmachari 1926, Two Neolithic Implements found in a Tank at Jamalpur (Monghyr), *JASB*, P. 136.

Brown, J. C. 1914, 'Grooved Stone Hammer from Assam and Similar Form in Eastern Asia,' *JPASB,* pp. 107-109.

___, 1917, *Catalogue of Prehistoric Antiquities in the Indian Museum,* Calcutta.

Brown, R. 1975, *Statistical Account of Manipur*, Delhi, P.4

Chakladar, H.C. 1941, 'The Prehistoric Culture of Bengal *MI.* IV, pp. 83- 105.

___, 1942, 'The Prehistoric Culture of Bengal', *MI.* XXII, pp. 140-162.

___, 1952, 'The Prehistoric Culture of Bengal', *MI,* XXXI, pp. 124-164.

Chakrabarti, Dilip. K. 1988, *Theoretical Issues in Indian Archaeology,* Delhi.

___, 1993, *Archaeology of Eastern India: Chhotanagpur Plateau and West Bengal,* Delhi.

___, 1999, *India: An Archaeological History,* Delhi.

Chakrabarti Dilip K and R.K. Chattopadhyay, 1984, The Lithic Industries of the Tarafeni and Bhairavbanki Valleys, West Bengal, *Calcutta Review,* II, pp.41 -63.

Chakrabarti, M. 1967, 'Occurence of Neolithic Implement in West Bengal', *BCRI,* pp 98-109.

Chakrabarti, M. and D. Mukherjee, 1971, *Indian Tribes,* Calcutta.

Chanda, S. and B.B Mukherjee, 1970, 'Radio Carbon Dating of Two Microfossiliferous Quaternary Deposit in and Around Calcutta' *SC,* XXXV, P. 275ff

Chanda, S. and Roy, P. 1990, 'Late Quaternary Vegetational History of Manipur', *JP,* XXVI, P. 51.

Chanda, R. P. 1976, *The Indo Aryan Race,* Rajsahi

Chang, K. C. 1967, 'Major Aspect of The Inter-Relationship of Archaeology and Ethonology', *CA,* VII, pp.227-228

Chang, K.C. 1974, 'Ancient Farmers in the Asain Tropics, Major Problems for Archaeological and Palaeo Environmental Investigation of South East Asia at The Earliest Neolithic Level,' In A.K. Ghosh (ed) *Perspective in Palaeo Anthropology,* Calcutta, pp. 36-43.

Chang, T.T. 1977, The Rice Culture' In *The Early History of Agriculture,* New York.

Chattopadhyay, B. D, G. Sengupta *et al* 2005 An Annotated Archalological Atlas of West Bengal, Vol.I.

Chatterjee, P. G. 1978, The Role and the Impact of Nature On Culture : Resplendent Results From Indian Neolithic In V. N. Misra, P. Bellwood (eds) *Recent Advances in Indo-Pacific Prehistory,* pp. 293-296.

Chatterjee, S. C. 1951-52, 'Discovery of Some Prehistoric Stone Tools in South Manbhum' *SC,* XVII (IV), P. 164.

Chatterjee, Suniti Kumar, 1951, *Kirata-Jana Krti: The Indo Mongoloid: Their Contribution to the History and Culture of India, RASB,* Calcutta.

Chowdhury, J.N. 1991, 'Prehistoric and Early Tribal Migrations in North-East India, In J.P.

Singh and G. Sengupta (eds) *Archaeology of North-Eastern India,* N.Delhi, P. 104.

Chaudhury, P.C. 1944, 'Neolithic Culture of Kamrupa' *JARS,* XI (II), P. 29.

Childe, V. G. 1936, *Man Makes Himself,* London.

___ 1952, *New Light on the Most Ancient East,* London.

1960, *What Happend in History,* London.

Cockburn, J. 1879, 'Stone Implement from Khasi Hills', *JASB,* pp. 107-109.

___ 1894, 'On Flint Implements from the Kon Ravines of South Mirzapure', *JRASB,* pp 21-27.

Condolle, Alphonse, 1884, *Origin of Cultivated Plants,* London, P.2.

Coon, C.S. 1963, *The Origin of Races,* New York.

Dalton, E. T. 1872, *Descriptive Ethnology of Bengal,* Calcutta.

Dani, A. H. 1960, *Prehistory and Protohistory of Eastern India,* Calcutta.

Das, B.M. 1961, *Introduction to Physical Anthropology,* Allahabad.

Das, S. R. 1968, *Stone Tools, History and Origin,* Calcutta.

Dash, R.N. 2008 *Archaeology, History and Culture of Orissa.*

Dasgupta, H.C. 1913, 'On Two Shouldered Implement from Assam' *JASB,* pp. 291-293

___ 1918, 'On Peculiar Polished Hammerstone from Singhbhum, Chhotanagpur, India', *IA,* XVII, pp. 135-136.

___, 1931, 'Bibliography of Prehistoric Indian Antiquities,' *JPASB.* pp. 1-96.

Dasgupta, P. C. 1963, *Archaeological Discoveries in West Bengal* Calcutta.

___, 1964, *The Excavations at Pandu Rajar Dhibi,* Calcutta.

___, 1966, *Exploring Bengal's past.* Directorate of Archaeology, Calcutta.

___, 1967, *Prgaitihasik Susunia, (Bengali),* Calcutta.

Datta, A. (ed), 1991, *Studies in Archaeology,* New Delhi.

___, 1992, *Neolithic Culture in West Bengal,* Delhi.

Datta Majumder, N. Some Prehistoric Finds in Orissa, *ISC,* 42nd Session, P. 321.

___,1972 , *Archaeological Seminar Paper,* Nagpur University.

Deo, S. B. (ed) , 1983, 'Ethnoarchaeology in India', *BDCRI,* Ll1, pp. 49-68.

Dhavalikar,M K, 1988 , *The First Farmars of the Deecan,* Pune. 1988.

Dhavalikar, M.K, H.D. Sankalia and Z. D. Ansari, 1988, *Excavation at Inamgaon,* Vol I, II, Pune.

Dhavalikar, M.K, 1995, Chalcolithic Architecture at Inamgaon and Walki : An Ethnoarchaeological Study' In B. Allchin(ed), *Living Tradition, Studies in Ethnoarchaeology of South Asia,* Oxford and 1BH Publishing, pp. 42-51.

Driem, G.V. 1998 Neolithic Correlates of Aneient Tibeto-Burman Migration in Blench, Spriggs (eds) *Archaeological Data and Linguistic Hypotheses.*

Fairservis, M. 1971 , *The Roots of Ancient India,* London.

Flannery, K.V. 1968, *Origin and Ecological Effects of Early Domestication in Iran and the Near-East*. Paper Presented at the Research Seminar in Archaeology and Related Subjects. *The Domestication and Exploitation of Plants and Animals,* London, pp. 18-19.

Fuchs, S. 1973 , *Aboriginal Tribe of India,* Calcutta.

Fuller, D.Q., Ling Qin, E. Harvey 2008 'A Critical Assesment of Early Agriculture in East Asia, with Emphasis on Dower Yangzte Rice Domestication' *Pragdhara* 18, pp. 17-52.

Gangte, T. S. 1993 , *Kukis of Manipur, A Historical Analysis,* New Delhi.

Ghosh, A. K. 1961, 'Prehistoric Studies in Eastern India', *SC,* XXVII, pp. 369-375.

Ghosh, N.C, 1962, 'Prehistoric Stone Implements from Midnapur' *SC*, XXVIII, pp. 338-339.

___, 1987, Palaeo Anthropology or Prehistory, Selection of Terminology in Indian Context', In B.M Pandey *etal* (eds), *Archaeology and History,* Delhi, pp. 3-10.

Ghosh, N.C. 1970, 'On The Neolithic Pottery of Eastern India' *JOI, XIX* pp. 333-339.

Godwin, Austen, H.H , 1875 , A Celt Found at Khasi Hills of Shillong,' *PASB*, P.158.

Goswami, K. G. 1948, *Excavation at Bangarh,* Calcutta.

Goswami, M.C. and Bhagabati, A. C. 1960, 'A Typological Study of Shouldered Celt from Rengchangiri', *JUG,* X, pp. 105-112.

Goswami, M.C. 1972, 'A Typological Study of Some Prehistoric Tools from Kameng District NEFA', *JASS,* XV (I), pp. 29-35.

Goswami, M. C and S. K. Roy, 1 975 , 'A Note on Tangsa Hoe, Arunachal Pradesh, *BDA,* IV, pp. 71-73.

Geldern, H. 1940, *Prehistoric Research in the Netherlands Indies in P. Honig (ed) Science and Scientist,* P. 138.

Gould, R. A. 'The Archaeologist as Ethnographer: A Case from the Western Desert of Australia', *WA,* III, pp. 143-177

Grignard, F.A, 1909, 'The Oraons and Mundas from the Time of Their Settlement in India' *Anthropos,* Tom IV, Heft 1, pp. 1-47.

Guha, B. S. 1931, 'Racial Affinities of People of India,' *C.O.I,* I (III) A.

___, 1937, 'An Outline of Ethnology of India', *Bull Ramkrishna Mission Institute of Culture.* Vol 1, P. 1-5.

Gupta, S.P. 1964, Remarks on 'Neolithic Problem in India' By B.K. Thapar in Misra and Mate (eds) *Indian Prehistory. Poona,* P. 100.

Gurdon, P. R.T. 1907, *The Khasis,* London

Haimendorf, C.V.F. 1945, 'The Problem of Megalithic Culture in the Middle India', *MI,* XXXVI (II) pp. 73-90

___, 1948 Notes on Stone Age in India', *MI,* XXXVIII, pp. 197-208

Hayden, B. 1981, 'Research and Development in the Stone Age : Technological Transitions Among Hunter Gatherers.' *CA,* IXXII, P. 519.

Hazarika, M 2006 'Neolithic Culture of North-East India : A Recent Perspective on the Origin of Pottery and Agriculture' *Ancient Asia,* I.

Hewitt, J. R, 1972, *The Ruling Races of Prehistoric Times in India, South Western Asia and Southern Europe,* Delhi.

Hodges, Henry, 1971 , *Technology in the Ancient World,* England.

___, 1976, *Artifacts; An Introduction to Early Material Technology,* London.

Hunday A. and S. Banerjee, 1967 , *'Geology and Mineral Resources of West Bengal', MGS1,* LXXXXVII, P. 302.

Hutton, J. H, 1921 , *The Angami Nagas,* London.

___, 1921 , *The Sema Nagas,* London.

___, 1924 , Two Celts from the Naga Hills,' *Man,* XXIV, (XV). pp. 20-24.

___, 1926, 'A Naga Hill Celt', *JASB* XXII, P. 133.

___, 1928, 'Prehistory of Assam', *MI,* VIII, pp. 228-232.

___, 1931, Ethnographic Notes by Various Hands, *COI* I (III)

Jain, R.C. 1970, *Ethnology of Ancient Bharat,* Varanasi.

Jarrige, Cathrine, Jean Francois, R.H. Meadow *et al* Mehrgarh field Reports 1974-84, from Neolithic Times to Indus Civilization.

Jaswal, I.J.S. 1976, 'Search for Early Man in North-Eastern India', *MI,* LXVI, pp. 276-277.

Jha,V, 1974 , From Tribe to Untouchables : The Case of Nisadas in R.S. Sharma (ed) *Indian Society, Historical Probing,* N. Delhi, P. 69,

Kosambi, D. D. 1967, 'Living Prehistory in India', *The American Review,* P. 29

Kramer, C. (ed), 1979, *Ethnoarchaeology: Implication of Ethnography for Archaeology,* New York.

Krishnaswamy, V. D. 1947, 'Stone Age India', *AI,* III, pp 11-58.

___, 1949 , 'Megalithic Types of South India', *AI,* V, P. 42ff.

___, 1953 'Progress In Prehistory', *AI,* IX, pp. 53-79.

___, 1960 , 'Neolithic Pattern in India', *AI,* 16, pp. 25-64.

Lal, B. B. 1958 , 'Birbhanpur: A Microlithic Site in the Damodar Valley. West Bengal', *AI,* XIV, pp. 4-98.

Leakey, L.S.B, 1975 , Working Stone, Bone and Wood in C. Singer (ed) *A History of Technology,* I, Oxford.

Le Mesuriar, H.P. 1861, 'Note on Twelve Stone Hatchets or Celts from the Neighbourhood of Jubbalpur', *PASB,* XXX, pp. 81-85.

Lubbock, J. 1867, 'Stone Tools of Assam', *Athenaeum,* 22nd June.

Lyall, Charles, 1908, *The Mikirs,* London.

Mackay, E. 1938 *Further Excavation at Mohenjodaro and Chanhudaro,* Delhi.

Manoj Kumar, 2003, An Ethnoarchaeological Study of Neolithic Culture of Northern and Eastern India (Unpublished Ph D Thesis)

Marshall, John, 1931, *Mohenjodaro and Indus Civilization,* London

Majumdar, D. N. 1958, *Races and Cultures of India,* Bombay.

Meadow, R.H, A.K. Patel 2002 From Mehrgarh to Harappa and Dholavira : Prehistoric Pastoralism in North-West South Asia through the Harappan Period.

Medhi, D. 1983, 'The Garo and Their Material Cutlure : A Study Based on Ethno-archaeologocial Approach', *ME,* VII, pp. 70-76.

Medlicott, H. B. 1875, 'A Stone Hatchet From Dibrugarh', *PASB,* P. 159.

Mills, J. P, 1929, *The Ao Nagas,* London.

Mill, J. P and Hutton, J. H, 1929, 'Ancient Monolith of North Cachhar', *JPASB,* XXV, pp. 295-297.

Misra, V. N. and Mate, M.S, 1964, *Indian Prehistory,* Poona.

Misra, V.N. and P. Bellwood (eds), 1978, *Recent Advances in Indo-Pacific Prehistory,* Delhi.

Mishra, P.K. (ed). 1997, *Comprehensive History and Culture of Orissa* I, II, N.Delhi.

Mitra, P. 1923 , *Prehistoric India,* Calcutta.

Mittre, V. and Gupta, H. 1970, 'Pollen Analytical Study of Quaternary Deposit in the Bengal Basin', *PB,* XIX (III), pp. 297- 306.

___, 1972, 'Neolithic Plant Economy at Chirand', *PB,* XXI (I), pp. 18-22.

Mohanty, P. 1986-87, 'Stone Age Research In Keonjhar District, Orissa' *Puratattva,* XVI, P. 54

Mukherjee, C. 1962, *The Santals,* Calcutta.

Mukherjee, T. 1 958, 'Neolithic Implement from Tamralipta' *IF,* I, pp. 42-44.

Nagar, M, 1 975 , 'Role of Ethnographic Evidence in the Reconstruction of Archaeological Data', *EA,* XXVIII (I). pp. 13-21.

___, 1983, 'Ethnoarchaeology of Bhimbhetka Region', *ME,* VII, pp. 61-70,

Narain, L.A. 1970, 'The Neolithic Settlement at Chirand, *JBRS* LVI

Oakley, K.P. 1 975, *Man the Tool Maker,* London.

Oldham, T. 1984, *Geology, Metrology and Ethnology of Meghalaya,* London, P. 22.

Orme Bryomy, 1981, *Anthropology for Archaeologist,* London

Paddayya, K. P. Bellwood 2002 South and South-East Asia in Archaeology, The Widening Debate.

Pal, J.N. 2009 Beginnings of Agriculture in the Middle Ganga Plain, Presidental Address. *U.P. History Congress.*

Pal, A. C. *etal.* 1996, *Prehistory of the Susunia Hill Complex and the Subarnarekha Valley, DAWB.*

Pandey, B.M and Chattopadhyay, B.D. (eds), 1987, *Archaeology and History,* I, II, Delhi,

Pande, G.C. (ed) 2009 The Dawn of India Civilization upto 600 B.C.

Peale, S.E, 1872, 'A Model Illustrating the Naga Method of Climbing Large Tress and A Celt', *PASB,* pp. 135-138.

Possehl, G.L. and P.C. Rissman 1992 *The Chronology of Prehistoric India:* From Earliest Times to the Iron Age. p.48.

Prasad, A. K. 1980-81, 'Excavation At Taradih', *Puratattva,* XII, pp. 138-139.

Rajguru, S. N. 1981, Geoclimatic Background of Prehistoric Assam in V.S. Srivastava (ed), *Cultural Contours of India,* New Delhi.

Rajguru, S.N, 1983, 'Problems of Late Pleistocene Aridity In India', *ME,* VIII, pp. 127-134.

Ramesh, N. R. 1986, 'Discovery of Stone Age Tools from Tripura and its Relevance to the Prehistory of South-East Asia', *G.S.M.* II, pp. 289-310.

Rami Reddy, 1991, *Neolithic and Post Neolithic Cultures,* New Delhi.

V.Randhawa, 1980 *A History of Agriculture In India*, Vol. 1, New Delhi.

Rao, S. N. 1973, 'Excavation at Sarutaru, A Neolithic Site of Assam', *ME*, I, pp. 1-9

Risley, H. H. 1915, *People of India,* Calcutta.

Rout S. P. 1969-70, 'Hand Book on Juang', *Adivasi,* XI, P. 1-53.

Roy, G.S. 1954, 'Microlithic Industry of Bongara', *MI,* XXXIV, pp. 16-19.

___, 1956, Implements of Neolithic Type in Bongara-Bhangat in Manbhoom, *MI,* XXXVI, pp. 49-55.

___, 1960 , 'Prehistory of Orissa', *MI*, XXXX (III), pp. 169-175.

Roy, G. S. and Ray, A. 1968 , 'A Technical Study of Some Prehistoric Potsherds' *MI,* pp. 147-154

Roy, H. M. 1986, *Savara, The Snake Charmar,* Calcutta.

Roy, Reba, 1987, *Ancient Settlement Pattern of Eastern India,* Calcutta.

Roy, S.C. 1912, *The Mundas and their Country,* Calcutta.

___, 1916, 'Note on Some Stone Implement Found in the Ranchi District', *JBORS,* II, pp. 61-77.

Roy Choudhury, P.C. 1965, *Bihar District Gazetteer, Santal Pargana,* Patna, P. 147

Sahi, M.D.N. 2010 'Neolithic Culture of the Ganga Valley : A New Perspective' in Tripathy, V. (ed) Archaeology of the Ganga Plain: A Paradigm shift. pp 94-122

Saraswat, K.S. 2004-05, Agricultural background of Early Farming Communities in the Middle Ganga Plain, *Pragdhara,* 15 pp. 145-177.

Sangma, M.S. 1991, Crafts and Technology of The Garos In J.P. Singh and G. Sengupta (eds) *Archaeology of North-Eastern India.* P. 304.

Sankalia, H.D. 1962, *Prehistory and Protohistory in India and Pakistan.* Poona.

___, 1962 *Indian Archaeology Today,* Bombay.

___, 1964, *Stone Age Tools,* Poona.

___, 1981, From History to Prehistory in Assam in V.S Srivastava(ed) *Cultural Contours of India,* New Delhi, pp. 81-94.

Sant, U. 1991, *Neolithic Settlement Pattern of North-Eastern and Northern India,* Delhi.

Sarkar, S.S. 1954, *The Aboriginal Races of India.* Calcutta.

Sauer, C.O. 1952, Agricultural Origin and Dispersal, *AGS,* New York pp. 20-22.

Semenov, S.A. 1957, *Prehistoric Technology,* London.

Sen, D. 1941, A Preliminery Note on Neolithic Typology of Chakradharpur, *PISC,* (28th Session) Pt. Ill

Sen, D. 1948, A Note on Some Celt and Chisel from West Bengal', *SC* X1V, pp. 252-253.

___, 1950 , 'A Celt Site in Singhbhum', *MI,* XXX(I), pp 1-14.

Sen, D and Chaturvedi, U. 1955, 'Further Finds of Stone Axes in Singbhum', *MI,* XXXV, (IV) pp. 303-314.

___, 1958, 'Further Studies In Singhbhum Neolithic Typology', *MI,* XXXVIII (III), pp. 176-185.

Sen, D. 1969, 'Profile of a Preliterate Culture in the Sanjai Valley' *JIAS IV,* pp. 17-29

Sen, D. 1975, Dungi: Profile of A Neolithic Site', *JIAS,* X, pp. 159-164.

Sengupta, G. Singh, J. (eds), 1991 *Archaeology of North-Eastern India,* New Delhi.

Settar, S, R. Settar (ed) 2002 *Prehistory : Archaeology of South Asia.*

Shaffer, J. G. 1981, *Prehistoric Baluchistan,* Delhi

___, 1986, *Cultural Development in the Eastern Punjab* in J. Jacobson(ed). *Studies in the Archaeology of India and Pakistan,* Delhi pp. 195-23 5.

Sharma, D.P. 2004 Prehistoric cultures of North-East India and South-East Asia : A comparative study in R.B. Foote Memorial National Seminar.

Sharma, C. Chauhan, M.S. Vegetation and Climate in Last Glacial Maximum in Darjeeling, Eastern Himalaya, *PISC,* (29th Session), P. 285

Sharma, G. R. *etal*, 1980, *Beginning of Agriculture (Epi Palaeolithic to Neolithic): Excavation at Chopani - Mando, Mahadaha and Mahagara,* Allahabad.

Sharma, G. R and J.D Clark (eds.), 1983, *Palaeo Evnvironment and Prehistory in the Middle Son Valley (Madhya Pradesh, North-Central India),* Allahabad.

Sharma, H.C. 1976, 'Stone Age Sites in Garo Hills, Meghalaya', *BDA,* V, pp. 82-92.

Sharma, H.C. and H.C Mahanta, 1988, Stone Age Cultures Around Selbal River, Western Garo Hill; *BDA*, II, pp. 29-36.

___, 1991-92, A Fresh Light on The Stone Age Culture of Rongram-Selbal Valley : Western Garo Hill, Meghalaya, *BDA*. Dibrugah University pp. 49-57.

Sharma, T.C, 1966, 'Researchs on the Prehistoric Archaeology of Assam', *JASS,* IX, pp. 1-11.

___, 1967, 'A Note on The Neolithic Pottery of Assam', *Man* II (1), pp. 126-128.

Sharma, T.C, 1974, Prehistory of Assam in D. Sen (ed) *Indian Anthropology Today,* Calcutta University, pp 33-70.

___, 1974, 'Recent Archaeological Discoveries in North-East India', *Puratattva* VII, pp. 17-19.

___, 1974, Prehistoric Archaeology in North East India its Problems and Prospects in B. M. Das (ed) *Contemporary Anthropological Research In Northeast India* ,pp. 147-155.

___, 1981, The Neolithic Pattern of North Eastern India in M.S Nagarja Rao (ed) *Madhu : Recent Researches in Indian Archaeology and Art History,* Delhi, P. 42.

Sharma, Y.D.1952, 'A New Stone Implement From Orissa', *MI,* XXXII, pp. 14-24.

Singh B. P. 2004, *Early Farming Communities of the Kaimur,* vol I, Jaipur.

Singh, G. 1990, 'Environmental Changes in Southern Asia During Holocene', *JP,* XXVI, P. 277.

Singh, K. S. 1994, *People of India,* Oxford University Press.

Singh, M. J. 1979, 'Prehistoric Archaeological Studies in Manipur', *Puratattva,* XII pp. 114-115.

Singh, O.K. 1972, 'Antiquities of Manipur' *Bull Manipur State Museum,* I, P. 1

Singh, M.M. 1964, 'Stone Age Economy of Eastern India' *JBORS,* L, pp. 875 -879

Singh, P.K. 1972, Cultural Succession of Prehistoric Culture in Central Orissa in K.L. Bhowmick (ed) *Culture of the Past* , Delhi, P. 95 ff

Singh, P. 1974, *Neolithic Culture of Western Asia.* London

___, 1997, *Excavations at Narhan,* Varanasi.

Singh, R.C.P., 1958, 'Prehistoric Bihar in Archaeological Perspective' *JBORS,* XLV(IIMV), pp. 198-200.

Singh, R. L. 1974, *India : A Regional Geography,* Varanasi.

Sinha, B.K. 1988-89, 'Excavation at Golbai Sasan', *Puratattva,* XXIII, pp. 48-50.

Sinha, B. P. 1979, *Archaeology and Art of India,* New Delhi. 1971

Sinha, B. N. 1971, *Geography of Orissa,* N. Delhi, P.5.

Sinha, B.P. and B.S. Verma, 1956, *Sonepur Excavation,* Directorate of Archaeology and Museum, Patna.

Sinha, C.P. (ed), 1998, *Facets of Indian Culture,* Patna.

Sinha, S. 1951, Discovery of Some Prehistoric Stone Tools in South Manbhum, *SC* VII.

Sinopoli, C. 1992, The Organization of Craft Production at Vijayanagara, South India', *AA*, LXXXX, pp.580-97.

Sonakia, A. 1984, 'The Skull Cap of Early Man and Associated Mammalian Fauna from Narmada Valley Alluvium, Hoshangabad Area, Madhya Pradesh (India)', *G.S.I Rec*, 113.Pt VI, pp. 159-171.

Spate, O.H.K, 1967, *India and Pakistan, A General Regional Geography*, London.

Srivastava, V.S. 1981, *Cultural Contours of India*. New Delhi.

Steel, E.H. 1870, 'Celt Found among Namsang Nagas', *PASB*, P. 267.

Stiles, Daniel, 1977, 'Ethonoarchaeology: A Discussion of Methods and Application', *Man*, *XII*, pp. 87-103.

Subbarao, B. 1958, *Personality of India: Pre and Protohistoric Foundation of India and Pakistan, Baroda*.

Thapar, B.K. 1974, 'Problem of Neolithic Culture in India: A Retrospect', *Puratattva*, VI, pp 61-65

___, 1983, Fresh Light on the Neolithic Culture of India in B.M. Pande and B.D. Chattopadhyay (eds) *Archaeology and History*, Delhi, pp. 247-254.

___, 1985 *Recent Archaeological Discoveries in India*, UNESCO, Japan.

Tewari, R, R. K. Srivastava *etal* 2008 'Early Farming at Lahuradewa.' *Pragdhara*, 18, pp. 346-73.

Theobold, W. 1875, 'Perforated Stone Implement From Kharakpur', *PASB*,pp. 102-103.

Trigger, B. 1971, 'Archaeology and Ecology', *WA*, III (II), pp 321-325.

Vavilov, N.I. 1949, 'The Origin, Variation, Community and Breeding of Cultivated Plants,' *CB*, XIII

Verma, A.K. 1994 , *Neolithic Culture of Eastern India. Delhi*

Verma, B.S. 1959, 'Archaeological Importance of Chirand', *JBORS*, XLV, (Pt I-IV).

___, 1970-71 'Excavation at Chirand', *Puratattva*, IV, pp. 19-23.

___, 2007 Chirand Excavation Report 1961-64 and 1967-1970, Directorate of Archaeology, Govt of Bihar, Patna

Verrier, Elwin, 1943, *The Aboriginals*, Oxford University Press.

Walsh, E.H.C, 1904, 'A Note on Stone Implement in Darjeeling District', *JASB*, LXXIII, (III), pp. 20-24.

Wheeler, R.E.M, 1947, 'Brahmagiri and Chandravalli: Megalithic and Other Cultures in Chital Durg District Mysore', *AI,* No. IV

White, L.A. 1959, *The Evolution of Culture,* New York, P. 284

Wright, G.S, 1971, 'Origin of Food Production in South-Western Asia,' *CA,* XII (IV-V), pp 447-477.

Wood, Mason, 1888, 'Note on Stone Object from a Neolithic Settlement Recently Discovered by W.H.P. Driver at Ranchi in Chhotanagpur', *JASB,* pp. 387-396.

___, 1889, 'Notice of a Neolithic Celt from Jashpur in Chhotanagpur', P. 254.

Worman, E. 1949, 'The Neolithic Problem in the Prehistory of India', *JWAS,* XXXIX (B), pp. 181-201.

Yasuda, Y. 2002 'Origin of Pottery and Agriculture in East Asia' in Yasuda (ed) *The Origins of Pottery and Agriculture*, Japan.

Index